TROPICAL FORESTS
OF OCEANIA

Anthropological Perspectives

TROPICAL FORESTS
OF OCEANIA

Anthropological Perspectives

EDITED BY JOSHUA A. BELL,
PAIGE WEST AND COLIN FILER

Australian
National
University

PRESS

ASIA-PACIFIC ENVIRONMENT MONOGRAPH 10

ANU PRESS

Published by ANU Press
The Australian National University
Acton ACT 2601, Australia
Email: anupress@anu.edu.au
This title is also available online at http://press.anu.edu.au

National Library of Australia Cataloguing-in-Publication entry

Title: Tropical forests of Oceania : anthropological perspectives
                / editors: Joshua A. Bell, Paige West,
                Colin Filer.

ISBN: 9781925022728 (paperback) 9781925022735 (ebook)

Series: Asia-Pacific environment monograph ; 10

Subjects: Rain forests--Oceania.
                Forest conservation--Oceania.
                Forests and forestry--Oceania.
                Human ecology--Oceania.

Other Creators/Contributors:
                Bell, Joshua A., editor.
                West, Paige, 1969- editor.
                Filer, Colin, editor.

Dewey Number: 577.34

Cover design and layout by ANU Press. Cover photograph, courtesy of Simon Foale, shows an oil palm estate in the making in East New Britain Province, Papua New Guinea.

Contents

Tables

Figures

Contributors

Joshua A. Bell is Curator of Globalization at the Smithsonian Institution's National Museum of Natural History.

Colin Filer is an Associate Professor in the Crawford School of Public Policy at The Australian National University.

Jennifer Gabriel is a Senior Research Officer in the Department of Anthropology, Archaeology and Sociology at James Cook University.

Jamon Alex Halvaksz is an Associate Professor in the Department of Anthropology at the University of Texas at San Antonio.

Edvard Hviding is a Professor of Anthropology and Founding Director of the Pacific Studies Research group at the University of Bergen.

Jerry K. Jacka is an Assistant Professor in the Department of Anthropology at the University of Colorado at Boulder.

Enock Kale is a Senior Scientist at the Papua New Guinea Institute of Biological Research.

Alexander Mawyer is an Assistant Professor in the Centre for Pacific Islands Studies at the University of Hawai'i at Manoa.

Paige West is a Professor of Anthropology in Barnard College and Columbia University.

Michael Wood is a Senior Lecturer in the Department of Anthropology, Archaeology and Sociology at James Cook University.

1. Introduction

JOSHUA A. BELL, PAIGE WEST AND COLIN FILER

Across the world indigenous rights activists argue that 'Land is Life!' This slogan appears on bumper stickers, T-shirts and posters at demonstrations against large-scale, capital-intensive forms of industrial development that result in indigenous peoples being dispossessed of their customary land-related rights and sovereignty. It indexes the fact that the systematic dispossession of such peoples from their lands has led to the current situation in which they are often living at the very margins of all international indicators of health and welfare. It also means much more than simply that land is important to life. For most indigenous groups globally, their lands and territories are inextricably tied to their ontological propositions about the world and their epistemic practices with regard to the world. Land is interwoven with history, memory, belief, practice and subjectivity. In this volume, we expand the discussion around indigenous peoples and land and argue that in many places in the larger forested nation-states of Oceania, forests are also life.

Native forests and their health are an enduring source of concern for the indigenous communities who have lived with them 'since time immemorial'. They are also a source of concern to many other actors or 'stakeholders' interested in their conservation or exploitation. Today, much of the news that we hear about forests is not good. There is an ongoing global debate about the pace of what is nowadays called deforestation and forest degradation, but there is widespread agreement that tropical forests are especially vulnerable (FAO 2010; FAO and ECJRC 2012; Hansen et al. 2013; Stibig et al. 2014; Kim et al. 2015). There is a parallel debate about the factors or drivers responsible for this loss in different parts of the world, and about the effects of this loss on local people's livelihoods. Yet there is general agreement that the loss of tropical forests is inextricably tied to a corresponding loss of biodiversity values, and this in turn has been linked to a decline in the health and welfare of the people living in and around these forests (Geist and Lambin 2002; Boucher et al. 2011). There is also a strong connection between the loss of biological diversity and the loss of social or cultural diversity, especially when the people affected by it are members of indigenous communities (Loh and Harmon 2005; Maffi 2005; Maffi and Woodley 2010; Gorenflo et al. 2012).

The contributors to this volume explore the ways in which indigenous people and other actors relate to the tropical forests of Oceania.[1] All but one of the chapters deal with this relationship in Papua New Guinea (PNG) and Solomon Islands, where the recent rate of forest loss has been most extreme. Calls for action to counter the process of deforestation have resulted in global monitoring plans, international and multilateral policy directives, and national-level laws and programs. While these are much needed, our concern is that the focus on large-scale solutions too often erases the very people whose lives are intertwined with their forests (Tsing 2005; West 2005). The tropical forests of Oceania, like those in other parts of the world, have always been constituted through human interaction with the natural environment (Groube 1989; Johns 1990; Kirch and Hunt 1997; Haberle et al. 2001; Bayliss-Smith et al. 2003; Balée 2006; Hope 2007; also Hviding, Chapter 3, this volume).

Our approach is grounded in political ecology, which means that we do not accept approaches to environmental degradation that seek to blame the rural poor for changes that have been precipitated by external forces, often by internationally financed extractive regimes. Based on a set of propositions about the ongoing nature of global processes of accumulation and dispossession, political ecology seeks to understand the complex relationships between changes at different social and ecological scales (Blaikie 1985; Blaikie and Brookfield 1987; Peet and Watts 1996; Zimmerer and Bassett 2003; Peet et al. 2011). Furthermore, political ecology is not only concerned with material forms of dispossession, but also with the set of semiotic processes that accompany them (Biersack 1999). Anthropologists working in this area have also been interested in the ecological foundations of human structures of thought and practice and the role of these structures in production of the natural world (West 2006; Halvaksz and Young-Leslie 2008; Jacka 2010). The present volume sits squarely in this tradition of scholarship. Forests are sites where social, ecological and economic relationships can be examined within larger political–economic and historical contexts (Peluso 1996; Ribot 1998). The chapters in this volume deal with a variety of topics already covered in different branches of the political ecology literature, including economic development and the commoditisation

---

1   The original drafts were presented at successive meetings of a session of the Association for Social Anthropology in Oceania between 2009 and 2011. The session was titled 'Forests of Oceania: Environmental Histories, Present Concerns and Future Possibilities', and was convened by Joshua Bell and Paige West.

of forests (Peet and Watts 1996; Bryant and Goodman 2004), the role of the state in resource dispossession (Bryant and Bailey 1997), the cultural politics of environmental transitions (Peluso 1992; Rocheleau and Ross 1995), the place of forests in the global commoditisation of nature (Castree 1995; West 2012), and the relationship between environmental conservation and environmental justice (Hecht and Cockburn 1989; Peluso 1993; West and Carrier 2004; West 2006).

The remaining native forests in the tropical regions of Oceania do not constitute a single type of ecosystem, but vary enormously with altitude and climate — from coastal mangrove systems at one extreme to high montane cloud forests at the other (Paijmans 1976; Gressitt 1982; Whitmore 1990; Hammermaster and Saunders 1995; Mueller-Dombois and Fosberg 1998; Bruijnzeel et al. 2011; Telnov 2011). In their incredible diversity, these forests have provided local communities with an equally wide variety of 'ecosystem services'. Perhaps most central has been the supply of food through the practice of shifting cultivation and the harvesting of tree crops, like the sago palm, that are more or less deliberately cultivated (Townsend et al. 1978; Clarke and Thaman 1997; Bourke and Allen 2009; Sillitoe 2010; Allen and Filer 2015). Although the centrality of this role has been diminished in some areas with increasing consumption of imported food commodities (Ohtsuka and Ulijaszek 2007), forests remain central to the livelihoods and sustenance strategies of many Pacific Island communities, even if forest resources are only used as a means of exchange for store-bought goods. Forests have also provided the raw materials for the construction of houses and canoes, both of which have been essential aspects of the different modalities of sociality, personhood, gender, space and time in Oceanic lifeworlds (Munn 1977; Bonnemaison 1986; Weiner 1991; Carucci 1995; Barlow and Lipset 1997; Rensel and Rodman 1997; Stasch 2003; Damon 2008).[2] By their very presence, forests root local cosmologies in the landscape and connect local communities to the places they inhabit (Nero 1992; West 2006; Crook 2007). They are also integral to the ways in which people inhabit the world, shaping people's sensory perceptions and aesthetics (Feld 1982; Gell 1995; Weiner 2001; Halvaksz 2003).

This volume explores the multiple ways in which the tropical forests of Oceania emerge in and through human action and imaginaries. Exploring the local dynamics within and around forests in the Pacific Islands region gives us insight into regional issues that have global resonance. Intertwined as they are with cosmological beliefs and livelihoods, as sites of biodiversity and Western desire, forests in this region have been and continue to be transformed by the

---

2   We use the past tense here because 'bush material' houses have been widely replaced by houses built with cement, milled timber and corrugated iron, while traditional sailing canoes have been widely replaced by aluminium dinghies or fibreglass 'banana boats'.

interaction of foreign and local entities. As one of the assemblages that form global capital, these forests are also sites of the various frictions that accompany these connections (Tsing 2005).

The present volume builds on the concerns of scholars and activists about the social and environmental impact of a boom in the export of raw logs from PNG and Solomon Islands in the 1990s (Gladman et al. 1996; Barlow and Winduo 1997; Filer 1997; Filer with Sekhran 1998; Dauvergne 2001; Hviding 2003; Forest Trends 2006; Wairiu 2007). These concerns were amplified by a raft of foreign investment in 'integrated conservation and development projects' in both countries following their ratification of the 1992 Convention on Biological Diversity (Kirsch 1997; McCallum and Sekhran 1997; van Helden 1998; Martin 1999; Hviding and Bayliss-Smith 2000; Filer 2004; Anderson 2005; West 2006; also West and Kale, Chapter 7, this volume). The logging boom was temporarily halted by the financial crisis of the late 1990s, and foreign funding for forest conservation projects began to decline at the same time. However, the volume of raw log exports from this part of Oceania has risen again over the course of the past decade, as the previous demand from Japan has been replaced by an equally strong demand from China (Dauvergne 1997; Filer 2013). In this recent period, the rising rate of timber extraction has not been matched by fresh investments in forest conservation, despite the global and national interest shown in the prospect of securing such investments for a reduction of carbon dioxide emissions from the process of deforestation and forest degradation (Melick 2010; Filer 2012; Filer and Wood 2012; Babon et al. 2014; also Filer, Chapter 8, Wood, Chapter 9, and Gabriel, Chapter 10, this volume).

Scholarly discussion of the political ecologies and economies of the Pacific Islands region has paid rather more attention to the impacts of mining than the impacts of logging (for example, Filer 1999; Rumsey and Weiner 2001; Kirsch 2006; West and Macintyre 2006; Bainton 2010; Golub 2014). This is partly because of the sheer scale and spectacle of big mining projects, but also because of the way that mining confronts and challenges the connection that communities have to their land. Indeed, the so-called 'selective logging' of native forests that is associated with the practice of 'sustainable forest management' does not typically involve the alienation of customary land rights or the dispossession of local landowners; it only involves the transfer of timber harvesting rights to logging companies for a fixed period of time (Taylor 1997; Filer with Sekhran 1998; Holzknecht 1998; Montagu 2002; Lea 2005). However, the latest logging boom in PNG has been associated with a process of forest clearance, or deforestation rather than forest degradation, that does entail a more complete form of dispossession because it is justified by plans to establish large-scale agricultural projects that are bound to displace subsistence livelihoods (Filer 2011, 2014; Winn 2012; Mousseau 2013). Instead of a cycle in which shifting cultivation (or selective logging) alternates with periods of forest regrowth or regeneration, forest ecosystems are permanently replaced by cash crop plantations. These could be timber plantations (see Hviding, Chapter 3, this volume), but the current crop of choice in Melanesia is oil palm (Filer 2010; Fraenkel et al. 2010; Allen 2012; Cramb and Curry 2012; Nelson et al. 2014).

Logging and mining can best be understood as two branches of extractive industry whose social and environmental effects are historically linked to booms and crashes in the value of specific commodities. Recent increases in the values of gold, copper, timber and oil palm are not necessarily associated with each other, but are each manifestations of the sort of commodity cycle through which the tropical forests of Oceania and their customary owners have come to be part of the world system over decades and even centuries. In most parts of the region, the ebbs and flows of commodity frontiers stretch back to the arrival of Europeans in the seventeenth century (Bennett 2000; Beinart and Hughes 2007; Newell 2010). In some areas, trade with Asia also has a long history (Swadling 1996; Spyer 2000; Dove 2011). And forests have not only been implicated in the export of specific commodities to Europe, Asia or America, but also in the importation of new products or commodities from these same regions, whether it be chainsaws in recent years or sweet potato since the seventeenth century (Ballard et al. 2005).

The history of the sandalwood trade illustrates these recursive loops and frictions. Needing a commodity to exchange for tea, Euro-American merchants discovered the value of the sweet-scented wood for the Chinese and quickly moved to exploit the groves found in Hawai'i in the early nineteenth century. These sources were exhausted by 1830 (Culliney 1988; Cuddihy and Stone 1990; Sahlins 1992), but the trade revived in the 1840s once new stands were located in the New Hebrides and New Caledonia (Shineberg 1967). During this second boom (1841–65), Melanesians, Polynesians and Europeans travelled through this area, establishing temporary camps or trading posts before shipping the wood to Australian or Asian ports.[3] These encounters were partly responsible for the creation of the Melanesian pidgin language (Romaine 1992: 34). If the conditions of extraction helped to generate new forms of sociality, they also helped to disrupt them. Within Hawai'i the sandalwood trade flooded the islands with foreign products that inspired contests of conspicuous consumption among the indigenous aristocracy (Sahlins 1992: 55–66). Prominent nobles put pressure on their tenants to gather sandalwood for the trade (ibid.: 57–82), the productivity of local agriculture declined, and the nobles fell deeper into debt.

---

3   Along the southern coast of what is now PNG, sandalwood exports did not begin until late in the nineteenth century, and continued almost to the time of independence in 1975 (Zieck 1970). Howcroft (1990) estimates that a total of 8,150 tonnes of sandalwood (180,000 trees) were sent to Singapore and China from Port Moresby before 1971.

Increased interaction with Euro-American sailors and merchants encouraged the transmission and outbreaks of new diseases which, combined with the harsh working conditions needed to procure the ever-diminishing supply of sandalwood, led to an estimated 25 per cent decrease in the indigenous population over a 20-year period (Sahlins 1992: 57; Kame'eleihiwa 1992: 80–2).

It is easy enough to understand the recent concern with the large-scale logging industry in the two Pacific Island countries that still boast extensive areas of unexploited native forest, but the history of the sandalwood trade should also alert us to the fact that specific tree species and much smaller areas of native forest have been implicated in profound forms of social and environmental change throughout the wider region (see Mawyer, Chapter 2, this volume). They were also implicated in such changes in PNG well before the large-scale log export industry emerged in the post-colonial period. For example, when Huli people in the central highlands saw Europeans cutting down the hoop pines that surrounded their sacred sites and using the wood as a building material, they concluded that white people must be related to their own ancestors who had originally planted the trees, and that European ownership of chainsaws and sawmills was itself a sign that Europeans owned the trees themselves (Schieffelin and Crittenden 1991: 271). This example shows that the associations of people with forests, or the frictions between them, cannot simply be understood as an effect of market forces.

Nevertheless, market forces have played a very significant part in the long-term transformation of these relationships. In the central highlands of PNG, the transformation can be seen through the widespread adoption of coffee as a smallholder cash crop (West 2012), just as it can also be seen through the previous adoption of sweet potato as a subsistence crop. Coffee is only one of several cash crops that have worked to transform the relationship between forest ecosystems or landscapes and indigenous communities in different parts of the region in different periods of recent history. Others include copra, cotton, cocoa, tea, sugar and rubber (Bennett 1987; McNeill 1988; Sahlins 1992; Lewis 1996; Ogan 1996; Kelly 1999; Tucker 2000). Whatever the specific form or scale of cultivation, these crops have jointly contributed to a process that Clarke and Thaman (1997) call 'agro-deforestation', which involves a reduction in the biological diversity of traditional forms of agroforestry. The recent oil palm boom in PNG and Solomon Islands is one phase in this long history, but the relationship of the oil palm industry to the process of deforestation has some novel features that relate to the revaluation of tropical forests as carbon sinks or biodiversity hotspots.

While large areas of native forest were cleared in the early years of the industry's development in PNG (Longayroux et al. 1972; Buchanan et al. 2008), the two companies that now export palm oil from PNG and Solomon Islands have both made a commitment to avoid any further deforestation as one of the conditions of having their products certified by the Roundtable on Sustainable Palm Oil (Nelson et al. 2010). This matters because the whole of their output is exported to Europe, where evidence of 'sustainability' attracts a better price. But the existing industry's reputation is now threatened by a raft of new proposals to fund the establishment of additional oil palm schemes from the profits to be made out of logging the native forest that they will replace, while depriving customary landowners of their land rights as well as their timber harvesting rights. Some of these so-called 'agroforestry' projects may be little more than logging projects disguised by agricultural development plans that will never be implemented, but even where a new oil palm estate does materialise, the operators evidently do not aim to prove their own sustainability. The industry has therefore come to be divided between companies that defend and companies that attack the same combination of indigenous rights and native forests (Filer 2013; Nelson et al. 2014).

The new generation of 'agroforestry' projects in PNG may well fit David Harvey's (2004) model of accumulation by dispossession, but the Roundtable on Sustainable Palm Oil illustrates a more recent transformation of tropical forest ecosystem services into a sort of commodity that companies can purchase as part of their 'social licence to operate'. Palm oil companies are not the only companies attracted to this market; logging companies can also seek to have their products certified if they think that this will raise their value for consumers (Atyi and Simula 2002; Rametsteiner and Simula 2003; Gulbrandsen 2004), while some mining and petroleum companies have shown an interest in purchasing carbon or biodiversity offsets to compensate for the environmental damage that they cause (Koziell and Omosa 2003; Maron et al. 2012; Virah-Sawmy et al. 2014). None of these corporate endeavours has yet had much effect on the relationship between people and forests in the Pacific Islands region, yet they reveal the capacity of capital to find new ways of doing business with this relationship.

This is not to deny the agency of indigenous people or local communities in the changes that actually happen or the way they are represented. Far from it, the chapters in this volume show how the peculiar histories of the region's many different types of forest can inform a challenge to policies or practices based on the assumption of a single historical trajectory. Together they show how Pacific Islanders continue to engage with the various processes at play in and around their forests, how the forests continue to generate social relationships, and how these relationships warrant the continued attention of anthropologists.

In Chapter 2, Alexander Mawyer discusses a process of rediscovery that is by no means uncommon in those parts of the Pacific Islands region where the indigenous population fell dramatically as a result of contact with Europeans and their diseases in the eighteenth and nineteenth centuries. When the population of Mangareva in French Polynesia fell below 500 in the 1870s, parts of the island that had once been densely settled and intensively cultivated began to revert to forest. The human population stabilised, recovered, and has recently grown quite rapidly. As a result, Mangarevans have begun to recolonise these areas of secondary forest and, in this process, have been rediscovering material evidence of former human occupation. This evidence has not only produced all sorts of curious entanglements with their own island's history; it has also become the subject of contemporary political debate about property, civility, sociality and identity.

In Chapter 3, Edvard Hviding shows how the indigenous population of New Georgia in the Solomon Islands was actively involved in the production of what later observers wrongly thought to be 'virgin' forests during the period in which the population of Mangareva fell so dramatically. Like Mawyer, Hviding contrasts the environmental transformations of the nineteenth century with those of the recent past, but in the case of New Georgia, the 'recolonisation' has taken the form of large-scale industrial logging and forest clearance, followed by the establishment of commercial tree plantations. What is most interesting about this transformation is that it has not involved the dispossession of local landowners by foreign investors, but has been organised by an indigenous social movement known as the Christian Fellowship Church. The leaders of this movement have a quite distinctive understanding of the relationship between their religious beliefs, their economic policies, and their position in a 'nation-state' that seems no more substantial than the 'virginity' of the native forests which they and their followers have removed from the local landscape.

In Chapters 4 and 5, Jamon Halvaksz and Jerry Jacka both examine the fate of the forest, or local people's relationship with local forests, in parts of PNG where mining has been a prominent and even dominant commercial activity during the colonial and post-colonial periods.

In Chapter 4, Halvaksz shows how Biangai people living in the upper part of the Bulolo River catchment in Morobe Province have responded to changes in the relationship between the logging and mining industries over a period of almost 90 years. While people from other parts of PNG were attracted to this area by periodic booms in large- and small-scale gold production, the Biangai people retained a close relationship to the towering pine trees that line the ridges of their landscape, and obtained more in the way of an income from the large-scale cultivation of these species. Their involvement in the social relations of compensation from the logging industry has more recently informed their response to the development of a new large-scale gold mine over the course of the past decade.

In Chapter 5, Jacka explores a somewhat different form of political ecology in the Porgera Valley in Enga Province, where mining has been the dominant commercial activity since gold was discovered by the Europeans who first contacted the Ipili people after World War II. The Ipili people themselves were quick to engage in small-scale mining activity, but their relationship to the local environment was profoundly transformed when a large-scale gold mine was developed in the 1990s and attracted a huge influx of people looking for a share of the economic opportunities it provided to local residents. Logging has never been a commercial activity in this area, but expansion of the mining industry has changed the way in which Ipili people make economic and cultural use of the diminishing amounts of different types of forest that have survived the process of industrial expansion and population growth.

In Chapter 6, Joshua Bell explores the entanglement of place, history and subjectivity in a very different part of PNG — the Purari Delta in Gulf Province. This is an area where the logging of native forests over the past two or three decades is part of a fractured history of efforts by outsiders to exploit a range of local resources, including cultural artefacts and human labour, since the area was first colonised in the 1880s. These experiences are central to local people's understanding of the constellation of relations between humans and non-humans that collectively comprise the delta's forests and waterways. Regional narratives of cultural loss and distant commodity abundance intersect with people's struggle to gain access to 'landowner benefits', including jobs and royalties, while dealing with the periodic incursion of police mobile squads searching for marijuana, home-made guns and stolen chainsaws. Overt and covert forms of violence are thus entangled with local people's perception of their connection with, or disconnection from, the nation-state and the wider world.

In Chapter 7, Paige West and Enock Kale review the history of one 'integrated conservation and development project' in the central highlands of PNG and show how it was unable to survive when some of the local villagers who were supposed to benefit from the 'development' component decided they preferred the prospect of large-scale mining or a petroleum project when this prospect began to materialise. While the apparent failure of this project can partly be explained by the reduction in funding from foreign sources, West and Kale argue that this and other similar conservation projects also failed because their neoliberal 'business models' placed them in the same game as their competitors in the business of 'development'. In other words, they were doomed to fail by their insistence on treating local people as rational economic actors who value forests or biodiversity because of their market potential. Nevertheless, West and Kale find some source of comfort in the fact that this 'business model' coexisted with a program of education and training that produced more substantial and enduring benefits for other Papua New Guineans.

The last three chapters deal with various aspects of the push to link forest conservation or forest management in PNG with international sources of finance for climate change mitigation by means of policies and projects that aim to reduce carbon dioxide emissions from the process of deforestation and forest degradation.

In Chapter 8, Colin Filer reviews the history of the one so-called REDD (Reducing Emissions from Deforestation and Forest Degradation) project that seems to have survived an international scandal about 'carbon cowboys' and a 'carbon cargo cult' that caused acute embarrassment for the PNG government in 2009. The April Salumei REDD project in East Sepik Province was adopted as a 'pilot project' by the PNG Forest Authority because the government had already acquired the right to turn a large area of forest into a logging concession but was unable to exercise this right because it was impossible to make binding agreements with all the different groups of customary landowners in the area. Nevertheless, a coalition of national 'stakeholders' was able to agree that a REDD project was the next best option, partly because it would not entail any further disturbance to the forest, but also because the question of how to distribute project benefits to local villagers was one whose answer could be postponed for some time — perhaps indefinitely.

In Chapter 9, Michael Wood tells the story of another REDD project that failed to survive the international scandal, partly because it achieved greater notoriety than all the rest. This was the Kamula Doso project in Western Province. Wood shows how the project proponent's engagement with the nascent (or perhaps illusory) 'national carbon economy' through the creation of exuberant images and narratives was way in excess of any 'rational' strategy for the acquisition of marketable property rights in forest carbon or the creation of a viable forest carbon market. His chapter therefore highlights the contingent role of such representations in a new world of connections between capital and conservation.

In Chapter 10, Jennifer Gabriel switches our attention from specific areas of tropical forest that are subject to specific forms of conservation or exploitation to the one big multinational company that has dominated PNG's logging industry for the past 25 years. She shows how Rimbunan Hijau has constructed an intricate web of power and meaning that aims to promote a corporate version of 'sustainable forest management' that can now be represented as an 'ethical market solution' to the problem of climate change. If Crater Mountain represents the failure of one neoliberal business model in the business of forest management, companies like Rimbunan Hijau have another type of neoliberal business model that continues to exert a lot more influence at different levels of political organisation, from the 'project' level to the national level and even the global level.

References

Allen, B. and C. Filer, 2015. 'Is the "Bogeyman" Real? Shifting Cultivation and the Forests, Papua New Guinea.' In M.F. Cairns (ed.), *Shifting Cultivation and Environmental Change: Indigenous People, Agriculture and Forest Conservation.* London: Routledge.

Allen, M.G., 2012. 'Informal Formalisation in a Hybrid Property Space: The Case of Smallholder Oil Palm Production in Solomon Islands.' *Asia Pacific Viewpoint* 53(3): 300–13.

Anderson, T., 2005. 'Challenging "Integrated Conservation and Development" in Papua New Guinea: The Bismarck-Ramu Group.' *Pacific Economic Bulletin* 20(1): 56–70.

Atyi, R.E and M. Simula, 2002. 'Forest Certification: Pending Challenges for Tropical Timber.' Yokohama: International Tropical Timber Organization (Technical Series 19).

Babon, A., D. McIntyre, G.Y. Gowae, C. Gallemore, R. Carmenta, M. Di Gregorio and M. Brockhaus, 2014. 'Advocacy Coalitions, REDD+, and Forest Governance in Papua New Guinea: How Likely Is Transformational Change?' *Ecology and Society* 19(3): 16.

Bainton, N.A., 2010. *The Lihir Destiny: Cultural Responses to Mining in Melanesia.* Canberra: ANU E Press (Asia-Pacific Environment Monograph 5).

Balée, W., 2006. 'The Research Program of Historical Ecology.' *Annual Review of Anthropology* 35: 75–98.

Ballard C., P. Brown, R.M. Bourke and T. Harwood (eds), 2005. *The Sweet Potato in Oceania: A Reappraisal.* Sydney: University of Sydney (Oceania Monograph 56). Pittsburgh: University of Pittsburgh, Department of Anthropology (Ethnology Monograph 19).

Barlow, K. and D. Lipset, 1997. 'Dialogics of Material Culture: Male and Female in Murik Outrigger Canoes.' *American Ethnologist* 24(1): 4–36.

Barlow, K. and S. Winduo (eds), 1997. *Logging in the Southwestern Pacific: Perspectives from Papua New Guinea, Solomon Islands, and Vanuatu.* Special Issue of *The Contemporary Pacific* 9(1).

Bayliss-Smith, T., E. Hviding and T.C. Whitmore, 2003. 'Rainforest Composition and Histories of Human Disturbance in Solomon Islands.' *Ambio* 32(5): 346–52.

Beinart, W. and L. Hughes, 2007. *Environment and Empire*. Oxford: Oxford University Press.

Bennett, J.A., 1987. *Wealth of the Solomons: A History of a Pacific Archipelago, 1800–1978*. Honolulu: University of Hawai'i Press.

Bennett, J.A., 2000. *Pacific Forest: A History of Resource Control and Contest in Solomon Islands, c. 1800–1997*. Cambridge: White Horse Press.

Biersack, A., 1999. 'Introduction: From the New Ecology to New Ecologies.' *American Anthropologist* 101(1): 5–18.

Blaikie, P., 1985. *The Political Economy of Soil Erosion in Developing Countries*. London: Longman.

Blaikie, P. and H. Brookfield, 1987. *Land Degradation and Society*. London: Methuen.

Bonnemaison, J., 1986. *Les Fondements d'une Identité: Territoire, Histoire et Société dans l'Archipel de Vanuatu (Mélanésie) – Livre I: L'Arbre et la Pirogue*. Bondy: ORSTOM.

Boucher, D., P. Elias, K. Lininger, C. May-Tobin, S. Roquemore and E. Saxon, 2011. *The Root of the Problem: What's Driving Tropical Deforestation Today?* Cambridge (MA): Union of Concerned Scientists.

Bourke, R.M. and B. Allen, 2009. 'Village Food Production Systems.' In R.M. Bourke and T. Harwood (eds), *Food and Agriculture in Papua New Guinea*. Canberra: ANU E Press.

Bruijnzeel, L.A., F.N. Scatena and L.S. Hamilton (eds), 2011. *Tropical Montane Cloud Forests: Science for Conservation and Management*. Cambridge: Cambridge University Press.

Bryant, R.L. and S. Bailey, 1997. *Third World Political Ecology*. London: Routledge.

Bryant, R.L. and M.K. Goodman, 2004. 'Consuming Narratives: The Political Ecology of "Alternative" Consumption.' *Transactions of the Institute of British Geographers* 29(3): 344–66.

Buchanan, G.M., S.H.M. Butchart, G. Dutson, J.D. Pilgrim, M.K. Steininger, K.D. Bishop and P. Mayaux, 2008. 'Using Remote Sensing to Inform Conservation Status Assessment: Estimates of Recent Deforestation Rates on New Britain and the Impacts upon Endemic Birds.' *Biological Conservation* 141(1): 56–66.

Carucci, L.M., 1995. 'Symbolic Imagery of Enewetak Sailing Canoes.' In R. Feinberg (ed.) *Seafaring in the Contemporary Pacific Islands: Studies in Continuity and Change*. Dekalb: Northern Illinois University Press.

Castree, N., 1995. 'The Nature of Produced Nature: Materiality and Knowledge Construction in Marxism.' *Antipode* 27(1): 12–48.

Clarke, W.C. and R.R. Thaman, 1997. 'Incremental Agroforestry: Enriching Pacific Landscapes.' *The Contemporary Pacific* 9(1): 121–48.

Cramb, R. and G.N. Curry, 2012. 'Oil Palm and Rural Livelihoods in the Asia-Pacific Region: An Overview.' *Asia Pacific Viewpoint* 53(3): 223–39.

Crook, T., 2007. *Anthropological Knowledge, Secrecy and Bolivip, Papua New Guinea: Exchanging Skin*. Oxford: Oxford University Press.

Cuddihy, L.W. and C.P. Stone, 1990. *Alteration of Native Hawaiian Vegetation: Effects of Humans, Their Activities and Introductions*. Honolulu: University of Hawai'i, Cooperative National Park Resource Studies Unit.

Culliney, J.L., 1988. *Islands in a Far Sea: Nature and Man in Hawaii*. San Francisco: Sierra Club.

Damon, F.H., 2008. 'On the Ideas of a Boat: From Forest Patches to Cybernetic Structures: The Outrigger Sailing Craft of the Eastern Kula Ring, Papua New Guinea.' In C. Sather and T. Kaartinen (eds), *Beyond the Horizon: Essays on Myth, History, Travel and Society*. Helsinki: Finnish Literature Society (Studia Fennica Anthropologica 2).

Dauvergne, P., 1997. *Shadows in the Forest: Japan and the Politics of Timber in Southeast Asia*. Cambridge (MA): MIT Press.

Dauvergne, P., 2001. *Loggers and Degradation in the Asia-Pacific: Corporations and Environmental Management*. Cambridge: Cambridge University Press.

Dove, M., 2011. *The Banana Tree at the Gate: The History of Marginal Peoples and Global Markets in Borneo*. New Haven (CT): Yale University Press.

FAO (Food and Agriculture Organization of the United Nations), 2010. 'Global Forest Resources Assessment 2010.' Rome: FAO (Forestry Paper 163).

FAO and ECJRC (Food and Agriculture Organization of the United Nations and European Commission Joint Research Centre), 2012. 'Global Forest Land-Use Change 1990–2005.' Rome: FAO (Forestry Paper 169).

Feld, S., 1982. *Sound and Sentiment: Birds, Weeping, Poetics, and Song in Kaluli Expression*. Philadelphia: University of Pennsylvania Press.

Filer, C., 2004. 'The Knowledge of Indigenous Desire: Disintegrating Conservation and Development in Papua New Guinea.' In A. Bicker, P. Sillitoe and J. Pottier (eds), *Development and Local Knowledge: New Approaches to Issues in Natural Resources Management, Conservation and Agriculture*. London: Routledge.

Filer, C., 2010. 'The Impacts of Rural Industry on the Native Forests of Papua New Guinea.' *Pacific Economic Bulletin* 25(3): 135–53.

Filer, C., 2011. 'New Land Grab in Papua New Guinea.' *Pacific Studies* 34(2/3): 269–94.

Filer, C., 2012. 'Why Green Grabs Don't Work in Papua New Guinea.' *Journal of Peasant Studies* 39(2): 599–617.

Filer, C., 2013. 'Asian Investment in the Rural Industries of Papua New Guinea: What's New and What's Not?' *Pacific Affairs* 86(2): 305–25.

Filer, C., 2014. 'The Double Movement of Immovable Property Rights in Papua New Guinea.' *Journal of Pacific History* 49(1): 76–94.

Filer, C. (ed.), 1997. *The Political Economy of Forest Management in Papua New Guinea*. London and Boroko: International Institute for Environment and Development and PNG National Research Institute (Monograph 32).

Filer, C. (ed.), 1999. *Dilemmas of Development: The Social and Economic Impact of the Porgera Gold Mine 1989–1994*. Canberra: Asia Pacific Press.

Filer, C., with N. Sekhran, 1998. *Loggers, Donors and Resource Owners*. London: International Institute for Environment and Development in association with the PNG National Research Institute.

Filer, C. and M. Wood, 2012. 'The Creation and Dissolution of Private Property in Forest Carbon: A Case Study from Papua New Guinea.' *Human Ecology* 40(5): 665–77.

Forest Trends, 2006. *Logging, Legality and Livelihoods in Papua New Guinea: Synthesis of Official Assessments of the Large-Scale Logging Industry*. Jakarta: Forest Trends.

Fraenkel, J., M. Allen and H. Brock, 2010. 'The Resumption of Palm-Oil Production on Guadalcanal's Northern Plains.' *Pacific Economic Bulletin* 25(1): 64–75.

Geist, H.J. and E.F. Lambin, 2002. 'Proximate Causes and Underlying Driving Forces of Tropical Deforestation.' *Bioscience* 52(2): 143–50.

Gell, A., 1995. 'The Language of the Forest: Landscape and Phonological Iconism in Umeda.' In E. Hirsch and M. O'Hanlon (eds), *The Anthropology of Landscape: Perspectives on Place and Space*. Oxford: Clarendon Press.

Gladman, D., D. Mowbray and J. Duguman (eds), 1996. *From Rio to Rai: Environment and Development in Papua New Guinea up to 2000 and Beyond – Volume 3: A Quarter of Next to Nothing*. Port Moresby: University of Papua New Guinea Press.

Golub, A., 2014. *Leviathans at the Gold Mine: Creating Indigenous and Corporate Actors in Papua New Guinea*. Durham (NC): Duke University Press.

Gorenflo, L.J., S. Romaine, R.A. Mittermaier and K. Walker-Painemilla, 2012. 'Co-occurrence of Linguistic and Biological Diversity in Biodiversity Hotspots and High Biodiversity Wilderness Areas.' *Proceedings of the National Academy of Sciences* 109(21): 8032–7.

Gressitt, J.L. (ed.), 1982. *Biogeography and Ecology of New Guinea*. The Hague: Dr W. Junk.

Groube, L., 1989. 'The Taming of the Rain Forests: A Model for Late Pleistocene Forest Exploitation in New Guinea.' In D.R. Harris and G.C. Hillman (eds), *Foraging and Farming: The Evolution of Plant Exploitation*. London: Unwin Hyman.

Gulbrandsen, L.H., 2004. 'Overlapping Public and Private Governance: Can Forest Certification Fill the Gaps in the Global Forest Regime?' *Global Environmental Politics* 4(2): 75–99.

Haberle, S.G., G.S. Hope and S. van der Kaars, 2001. 'Biomass Burning in Indonesia and Papua New Guinea: Natural and Human Induced Fire Events in the Fossil Record.' *Palaeogeography Palaeoclimatology Palaeoecology* 171(3/4): 259–68.

Halvaksz, J., 2003. 'Singing about the Land among the Biangai.' *Oceania* 7(3): 153–69.

Halvaksz, J. and H.E. Young-Leslie, 2008. 'Thinking Ecographically: Places, Ecographers and Environmentalism.' *Nature and Culture* 3(2): 183–205.

Hammermaster, E.T. and J.C. Saunders, 1995. 'Forest Resources and Vegetation Mapping of Papua New Guinea.' Canberra: Australian Agency for International Development, PNG Resource Information System (Publication 4).

Hansen, M.C., P.V. Potapov, R. Moore, M. Hancher, S.A. Turubanova, A. Tyukavina, D. Thau, S.V. Stehman, S.J. Goetz, T.R. Loveland, A. Kommareddy, A. Egorov, L. Chini, C.O. Justice and J.R.G. Townshend, 2013. 'High-Resolution Global Maps of 21st-Century Forest Cover Change.' *Science* 342: 850–3.

Harvey, D., 2004. 'The "New" Imperialism: Accumulation by Dispossession.' *Socialist Register* 2004: 63–87.

Hecht, S. and A. Cockburn, 1989. *The Fate of the Forest: Developers, Destroyers, and Defenders of the Amazon.* London: Verso.

Holzknecht, H., 1998. 'Problems of Articulation and Representation in Resource Development: The Case of Forestry in Papua New Guinea.' *Anthropological Forum* 7(4): 549–73.

Hope, G.S., 2007. 'The History of Human Impact on New Guinea.' In A.J. Marshall and B.M. Beehler (eds), *The Ecology of Papua.* Singapore: Periplus Editions.

Howcroft, N.S.H., 1990. 'A Review of Papua New Guinea Sandalwood.' Lae: PNG Department of Forests.

Hviding, E., 2003. 'Contested Rainforests, NGOs, and Projects of Desire in Solomon Islands.' *International Social Science Journal* 178: 539–54.

Hviding, E. and T. Bayliss-Smith, 2000. *Islands of Rainforest: Agroforestry, Logging and Eco-Tourism in Solomon Islands.* Aldershot: Ashgate.

Jacka, J.K., 2010. 'The Spirits of Conservation: Ecology, Christianity, and Resource Management in Highlands Papua New Guinea.' *Journal for the Study of Religion, Nature, and Culture* 4(1): 24–47.

Johns, R.J., 1990. 'The Illusionary Concept of the Climax.' In P. Baas, K. Kalkman and R. Geesink (eds), *The Plant Diversity of Malesia.* Dordrecht: Kluwer Academic Publishers.

Kame'eleihiwa, L., 1992. *Native Land and Foreign Desires: Pehea Lā E Pono Ai? How Shall We Live in Harmony?* Honolulu: Bishop Museum Press.

Kelly, J., 1999. 'The Other Leviathans: Corporate Investment and the Construction of a Sugar Colony.' In P. Ahluwadia, B. Ashcroft and R. Knight (eds), *White and Deadly: Sugar and Colonialism.* Cammack (NY): Nova Science.

Kim, D.-H., J.O. Sexton and J.R. Townshend, 2015. 'Accelerated Deforestation in the Humid Tropics from the 1990s to the 2000s.' *Geophysical Research Letters* DOI 10.1002/2014GL062777.

Kirch, P.V. and T.L. Hunt (eds), 1997. *Historical Ecology in the Pacific Islands: Prehistoric Environmental and Landscape Change*. New Haven (CT): Yale University Press.

Kirsch, S., 1997. 'Regional Dynamics and Conservation in Papua New Guinea: The Lakekamu River Basin Project.' *The Contemporary Pacific* 9(1): 97–121.

Kirsch, S., 2006. *Reverse Anthropology: Indigenous Analysis of Social and Environmental Relations in New Guinea*. Stanford (CA): Stanford University Press.

Koziell, I. and E. Omosa, 2003. 'Room to Manoeuvre? Mining, Biodiversity and Protected Areas.' London: International Institute for Environment and Development.

Lea, D., 2005. 'The PNG Forest Industry, Incorporated Entities and Environmental Protection.' *Pacific Economic Bulletin* 20(1): 168–77.

Lewis, D.C., 1996. *The Plantation Dream: Developing British New Guinea and Papua, 1884–1942*. Canberra: Journal of Pacific History.

Loh, J. and D. Harmon, 2005. 'A Global Index of Biocultural Diversity.' *Ecological Indicators* 5(3): 231–41.

Longayroux, J.P., T. Fleming, A. Ploeg, R.T. Shand, W.F. Straatmans and W. Jonas, 1972. 'Hoskins Development: The Role of Oil Palm and Timber.' Boroko: The Australian National University, New Guinea Research Unit (Bulletin 49).

Maffi, L., 2005. 'Linguistic, Cultural, and Biological Diversity.' *Annual Review of Anthropology* 29: 599–617.

Maffi, L. and E. Woodley (eds), 2010. *Biocultural Diversity Conservation: A Global Sourcebook*. London: Earthscan.

Maron, M., R.J. Hobbs, A. Moilanen, J.W. Matthews, K. Christie, T.A. Gardner, D.A. Keith, D.B. Lindenmayer and C.A. McAlpine, 2012. 'Faustian Bargains? Restoration Realities in the Context of Biodiversity Offset Policies.' *Biological Conservation* 155: 141–8.

Martin, R.E., 1999. *Integrating Conservation and Development in a Papua New Guinean Community*. Melbourne: Monash University, School of Geography and Environmental Science (Publications in Geography 52).

McCallum, R. and N. Sekhran, 1997. *Race for the Rainforest: Evaluating Lessons from an Integrated Conservation and Development 'Experiment' in New Ireland, Papua New Guinea*. Port Moresby: PNG Biodiversity Conservation and Resource Management Programme.

McNeill, J., 1988. 'From Magellan to Miti: Pacific Rim Economies and Pacific Island Ecologies since 1521.' In S. Miller, A.J.H. Latham and D.O. Flynn (eds), *Studies in the Economic History of the Pacific Rim*. London: Routledge.

Melick, D., 2010. 'Credibility of REDD and Experiences from Papua New Guinea.' *Conservation Biology* 24(2): 359–61.

Montagu, A.S., 2002. 'Forest Planning and Management in Papua New Guinea, 1884 to 1995: A Political Ecological Analysis.' *Planning Perspectives* 17(1): 21–40.

Mousseau, F., 2013. 'On Our Land: Modern Land Grabs Reversing Independence in Papua New Guinea.' Oakland (CA): Oakland Institute in collaboration with Pacific Network on Globalisation.

Mueller-Dombois, D. and F.R. Fosberg, 1998. *Vegetation of the Tropical Pacific Islands*. New York: Springer.

Munn, N., 1977. 'The Spatiotemporal Transformations of Gawa Canoes.' *Journal de la Société des Océanistes* 33(1): 39–51.

Nelson, P.N., J. Gabriel, C. Filer, M. Banabas, J.A. Sayer, G.N. Curry, G. Koczberski and O. Venter, 2014. 'Oil Palm and Deforestation in Papua New Guinea.' *Conservation Letters* 7(3): 188–95.

Nelson, P.N., M.J. Webb, I. Orrell, H. van Rees, M. Banabas, S. Berthelsen, M. Sheaves, F. Bakani, O. Pukam, M. Hoare, W. Griffiths, G. King, P. Carberry, R. Pipai, A. McNeill, P. Meekers, S. Lord, J. Butler, T. Pattison, J. Armour and C. Dewhurst, 2010. 'Environmental Sustainability of Oil Palm Cultivation in Papua New Guinea.' Canberra: Australian Centre for International Agricultural Research (Technical Report 75).

Nero, K.L., 1992. 'The Breadfruit Tree Story: Mythological Transformations in Palauan Politics.' *Pacific Studies* 15(4): 235–60.

Newell, J., 2010. *Trading Nature: Tahitians, Europeans, and Ecological Exchange*. Honolulu: University of Hawai'i Press.

Ogan, E., 1996. 'Copra Came before Copper: The Nasioi of Bougainville and Plantation Colonialism, 1902–1964.' *Pacific Studies* 19(1): 31–50.

Ohtsuka, R. and S.J. Ulijaszek (eds), 2007. *Health Change in the Asia-Pacific Region: Biocultural and Epidemiological Approaches*. Cambridge: Cambridge University Press.

Paijmans, K. (ed.), 1976. *New Guinea Vegetation.* Canberra: Commonwealth Scientific Industrial Research Organization and Australian National University Press.

Peet, R., P. Robbins and M. Watts (eds), 2011. *Global Political Ecology.* New York: Routledge.

Peet, R. and M. Watts (eds), 1996. *Liberation Ecologies: Environment, Development, Social Movements.* London: Routledge.

Peluso, N.L., 1992. *Rich Forests, Poor People: Resource Control and Resistance in Java.* Berkeley: University of California Press.

Peluso, N.L., 1993. 'Coercing Conservation? The Politics of State Resource Control.' *Global Environmental Change* 3(2): 199–217.

Peluso, N.L., 1996. 'Fruit Trees and Family Trees in an Anthropogenic Forest: Ethics of Access, Property Zones, and Environmental Change in Indonesia.' *Comparative Studies in Society and History* 38(3): 510–48.

Rametsteiner, E. and M. Simula, 2003. 'Forest Certification – an Instrument to Promote Sustainable Forest Management?' *Journal of Environmental Management* 67: 87–98.

Rensel, J. and M. Rodman (eds), 1997. *Home in the Islands: Housing and Social Change in the Pacific.* Honolulu: University of Hawai'i Press.

Ribot, J.C., 1998. 'Theorizing Access: Forest Profits along Senegal's Charcoal Commodity Chain.' *Development and Change* 29(2): 307–41.

Rocheleau, D. and L. Ross, 1995. 'Trees as Tools, Trees as Text: Struggle over Resources in Zambrana-Chacuey, Dominican Republic.' *Antipode* 27(4): 407–28.

Romaine, S., 1992. *Language, Education, and Development: Urban and Rural Tok Pisin in Papua New Guinea.* Oxford: Clarendon Press.

Rumsey, A. and J. Weiner (eds), 2001. *Mining and Indigenous Lifeworlds in Australia and Papua New Guinea.* Adelaide: Crawford House.

Sahlins, M., 1992. *Anahulu: The Anthropology of History in the Kingdom of Hawaii – Volume 1: Historical Ethnography.* Chicago: University of Chicago Press.

Schieffelin, E.L. and R. Crittenden (eds), 1991. *Like People You See in a Dream: First Contact in Six Papuan Societies.* Stanford (CA): Stanford University Press.

Shineberg, D., 1967. *They Came for Sandalwood: A Study of the Sandalwood Trade in the South-West Pacific 1830–1865*. Melbourne: Melbourne University Press.

Sillitoe, P., 2010. *From Land to Mouth: The Agricultural 'Economy' of the Wola of the New Guinea Highlands*. New Haven (CT): Yale University Press.

Spyer, P., 2000. *The Memory of Trade: Modernity's Entanglements on an Eastern Indonesian Island*. Durham (NC): Duke University Press.

Stasch, R., 2003. 'The Semiotics of World-Making in Korowai Feast Longhouses.' *Language & Communication* 23(3/4): 359–83.

Stibig, H-J., F. Achard, S. Carboni, R. Raši and J. Miettinen, 2014. 'Change in Tropical Forest Cover of Southeast Asia from 1990 to 2010.' *Biogeosciences* 11: 247–58.

Swadling, P., 1996. *Plumes from Paradise: Trade Cycles in Outer Southeast Asia and Their Impact on New Guinea and Nearby Islands until 1920*. Boroko: PNG National Museum in association with Robert Brown & Associates.

Taylor, R., 1997. 'The State versus Custom: Regulating Papua New Guinea's Timber Industry.' In C. Filer (ed.), *The Political Economy of Forest Management in Papua New Guinea*. London and Boroko: International Institute for Environment and Development and PNG National Research Institute (Monograph 32).

Telnov, D. (ed.), 2011. *Biodiversity, Biogeography and Nature Conservation in Wallacea and New Guinea*. Riga: Entomological Society of Latvia.

Townsend, P.K., K.R. Ruddle, D. Johnson and J. Rees, 1978. *Palm Sago: A Tropical Starch from Marginal Lands*. Honolulu: University of Hawai'i Press.

Tsing, A.L., 2005. *Friction: An Ethnography of Global Connection*. Princeton (NJ): Princeton University Press.

Tucker, R.P., 2000. *Insatiable Appetite: The United States and the Ecological Degradation of the Tropical World*. Berkeley: University of California Press.

van Helden, F., 1998. *Between Cash and Conviction: The Social Context of the Bismarck-Ramu Integrated Conservation and Development Project*. Boroko: PNG National Research Institute (Monograph 33).

Virah-Sawmy, M., J. Ebeling and R. Taplin, 2014. 'Mining and Biodiversity Offsets: A Transparent and Science-Based Approach to Measure "No-Net-Loss".' *Journal of Environmental Management* 143: 61–70.

Wairiu, M., 2007. 'History of the Forestry Industry in Solomon Islands.' *Journal of Pacific History* 42(2): 233–46.

Weiner, J.F., 1991. *The Empty Place: Poetry, Space, and Being among the Foi of Papua New Guinea*. Bloomington: Indiana University Press.

Weiner, J.F., 2001. *Tree Leaf Talk: A Heideggerian Anthropology*. Oxford: Berg.

West, P., 2005. 'Translation, Value, and Space: Theorizing an Ethnographic and Engaged Environmental Anthropology.' *American Anthropologist* 107(4): 632–42.

West, P., 2006. *Conservation Is Our Government Now: The Politics of Ecology in Papua New Guinea*. Durham (NC): Duke University Press.

West, P., 2012. *From Modern Production to Imagined Primitive: The Social World of Coffee from Papua New Guinea*. Durham (NC): Duke University Press.

West, P. and J.G. Carrier, 2004. 'Ecotourism and Authenticity: Getting Away from It All?' *Current Anthropology* 45(4): 483–98.

West, P. and M. Macintyre (eds), 2006. *Melanesian Mining Modernities: Past, Present, and Future*. Special Issue of *The Contemporary Pacific* 18(2).

Whitmore, T.C., 1990. *An Introduction to Tropical Rain Forests*. Oxford: Clarendon Press.

Winn, P., 2012. 'Up for Grabs: Millions of Hectares of Customary Land in PNG Stolen for Logging.' Sydney: Greenpeace Australia Pacific.

Zieck, J., 1970. 'Minor Forest Products: Sandalwood (and Rattan) in Rigo Sub District.' Hohola: Territory of PNG, Department of Forests.

Zimmerer, K.S. and T.J. Bassett (eds), 2003. *Political Ecology: An Integrative Approach to Geography and Environment-Development Studies*. New York: Guilford Press.

2. Wildlands, Deserted Bays and Other Bushy Metaphors of Pacific Place

ALEXANDER MAWYER

> It is not only in the modern imagination that forests cast their shadow of primeval antiquity; from the beginning they appeared to our ancestors as archaic, as antecedent to the human world (Harrison 1992: 1).

In French Polynesia's Gambier Islands, contemporary Mangarevans have a disconcerting relationship with their islands' imputable wilds and forested spaces. Traditionally and generically referred to as the *vao*, these nature spaces were radically altered by a series of transformative ecosocial projects undertaken by Catholic missionaries beginning in the 1830s. These islands concurrently experienced a massive depopulation in the following decades. As a result, many previously densely inhabited and cultivated bays, and the slopes upland from them on Mangareva and other islands in the Gambier archipelago, acquired a shaggy veneer in categories of local understanding as no-longer and not-quite domestic terrain. A century later, regional government-sponsored forestry efforts only enhanced and amplified the landscape's transformation. If this is a pattern familiar in anthropological literature across the Pacific, it is nowhere more evident than in some of the Polynesian island groups of the eastern Pacific, where the demographic, social and cultural consequences of European contact were particularly stark (Dening 1988; Kirch and Rallu 2007). I suggest that the unfolding of this history at the intersection of these islands' nature spaces and its legacy in various aspects of local experience and understanding has been a process of de- and fraught re-domestication that casts a peculiar shadow on such ecosocial conceptions as 'forest' or 'wilds'. This chapter thus contributes to the observation in the works collected here that the forest in Oceania is not only an object of contemporary resource management, exploitation and human processes playing out over the long timescales of slow-life trees and their mature ecosystems, but also a historically inflected site of cultural being, social identity and the politics of the everyday with dynamic features over time.

Although it is hoped that some points raised here may be relevant to considerations of the forest across the region, my focus is solidly centred in the French *pays d'outre-mer* of eastern Polynesia and particularly in their southernmost group, the Gambier Islands. French Polynesia is an aqueous quasi-autonomous French dependency in the south-eastern Pacific. Its territory comprises five island groups, each distinct by geography, language, culture and history. The Society Islands are the regional centre, comparatively favoured by land area and population. Most of the Society Islands are high and lush with large lagoons that separate their main islands from the barrier reefs that encircle them. The regional capital, Pape'ete, on Tahiti, is in the windward south-east of the group and is now home to the vast majority of the territory's population, currently growing upwards of 270,000 (ISPF 2011). The Tuamotu Archipelago, previously called The Dangerous for its treacherous currents and submerged *recifs*, is the largest group by number of islands, nearly 80 atolls. Seen from the sky, these are circles of low coral islets on reefs that define large lagoons. The widely dispersed Austral Islands to the far south of the region are partly out of the tropics and famous for their artisanal productions and pre-contact ridgetop fortifications. Further to the north-east, and north of the Tuamotu, lie the well-known Marquesas, made famous by Melville's romantic tale of a sojourn in Taipivai Valley on Nuku Hiva and Gauguin's fauvist escapism on Hiva Oa. These are high islands with sheer and rugged coastlines but without the coral reef structures found elsewhere in French Polynesia — sometimes argued to have encouraged a high degree of isolationism and inter-valley conflict. To the south-east of the Tuamotu are the Gambier Islands, once an articulated volcanic structure in ages past, now several island fragments of which Mangareva is the largest within a single large reef structure notable for numerous low coral islets at different points along the reef.

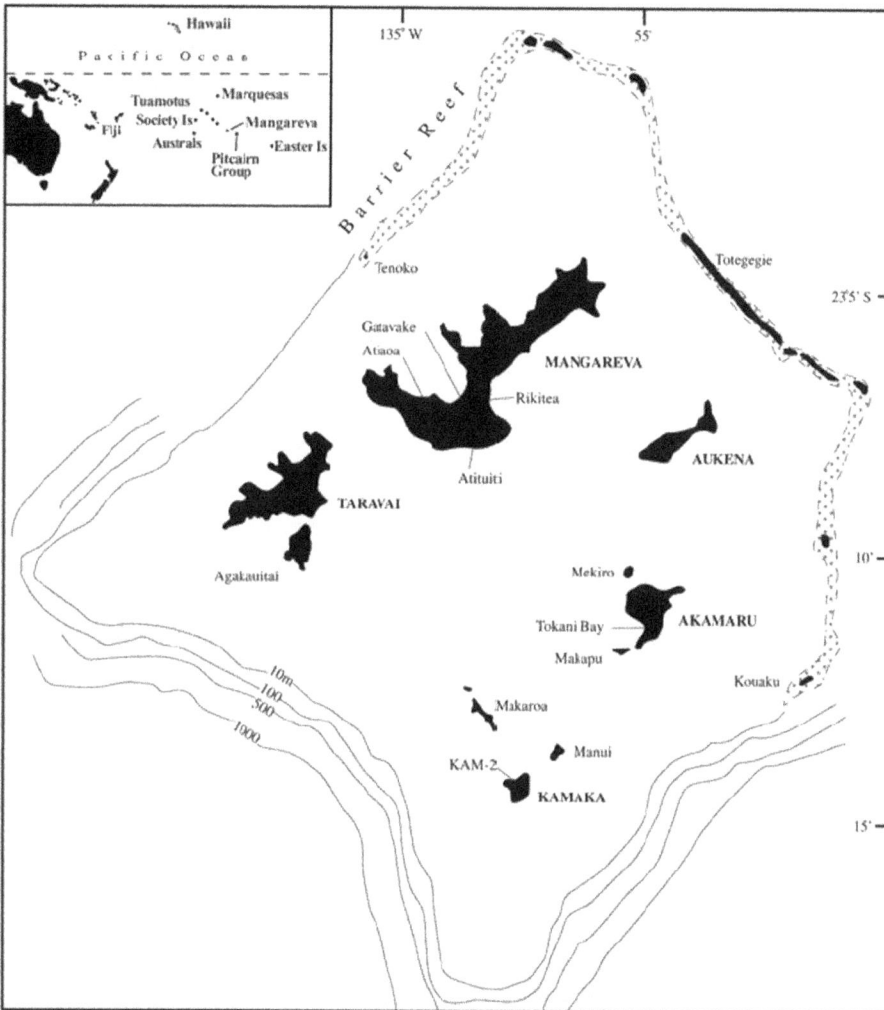

Figure 2.1 The Gambier Islands.
Source: Anderson et al. (2003: 120).

The Gambier rise from the oceanic deeps 1,700 kilometres south-east of the Society Islands. They are eight high islands amounting to 27 square kilometres of land surface and numerous *motu* (sand islands) or islets encircling a 90-kilometre barrier reef enclosing a vibrant coral lagoon. One notes that even such small islands as these have non-trivially varied micro-ecologies. Terrains including high grassed peaks, upland pine forests, complex traditional and neo-traditional downslope forests and highly varied near-shore house gardens and plantations. *Motu* on the fringing reef evidence well-established coconut plantations. Before contact with Europeans, all the islands of the chain were densely occupied, as were the sandy barrier islands, relative to their limited freshwater lens and the outlying atoll Temoe to the east of the group (Buck 1938; Laval 1938). Only the three islands of Aukena, Akamaru and Mangareva were inhabited between 1999 and 2009, the years between my first field visit and my last. Other islands in the group were home to a few families until the 1960s, while Aukena and Akamaru are home to only a few families today.

Today, however, the Gambier's population is burgeoning and beginning to push back into these formally (un)inhabited spaces, yielding encounters that can be both surprising and disconcerting in their historical inflections as families — whether youthful or mature couples — clear land for the first time in many decades, or in a century. What are contemporary wilds can turn out to have been someone's *kaiga* (domesticated lands) containing a house platform, boundary-marking sitting stone, or stonework structure of unknown utility. Apparent gnarly woods can also contain one or more charming French provincial quarried stone houses from the height of the Catholic instauration in the mid-nineteenth century, long abandoned and often forgotten, alongside stone ruins of earlier cultural complexes. Encounters with the material facts of historical displacements, discontinuities and gaps, lead contemporary Mangarevan landowners to ponder uncanny questions of historic loss over several different periods.

Moreover, the signs and sediments of history (Douglas 2009; Jolly 2009) are not limited to the artefactual or the architectural. Plant cultures too, often literally beneath one's feet should one walk along contemporary roads or into house gardens in any direction, or apperceived in the increasingly resplendent verdure of the quotidian, also reflect the struggles and counter-struggles of individuals in the receding past to make and remake the world-space of their home islands from the era of early European contacts, through the height of the colonial, and into the present through cultivation and gardening practices. This is visible not only in domestic landscapes surrounding today's *'akairiga* (cohabiting family) homes; the contemporary wilds are also woven with cultures of the past, floral as well as material.

For instance, here and there in splendid isolation across the now wooded and bushy spaces of the high islands, often near visible remains of past dwellings or other structures, one finds aged coffee stands, a raspberry tangle deep in woods, soaring breadfruits, small groves of mature lychee trees, or surprisingly durable wild orange trees among other more modest food plants from mangoes to starfruits to tamarinds, many now evidently entangled in their use-rights. When these signs of past domestication are found on the back (*tua*) side of Mangareva, or in the bays of the smaller islands in an archipelago tragically depopulated a century earlier, their presence can be more than striking. With Douglas (2009) one notes that they offer enduring, hardy signs of the islands' histories and perhaps counter-histories at the intersection of plant and human cultures in the eastern Pacific. Individual fruit trees, berry tangles or a small stand of citrus or coffee in the midst of bush stand as silent sentinels of a prior social and ecological order.

Drawing on fieldwork in the Gambier since my own first arrival in these islands in 1999, when I participated in a land-clearing and restoration party organised by the Catholic Church on the then-uninhabited island of Akamaru, to working with two different households' land clearing, planting and home-planning practices in the summer of 2009, I suggest that Mangarevan understandings of the relationship of de-domesticated terrains and lands, *te vao* (the wilds), to domestic spaces, *punui* (habitation centres, towns), is in a period of significant flux if not confusion. An ambiguous quality of the landscape reveals something about the character of contemporary social life and culture in the Gambier, the active roles of the past in the present, and implications for the consideration of forest and other ontological objectifications of nature in the eastern Pacific.

By way of foreshadowing this chapter's subsequent sections, this work joins others that have articulated a profoundly warranted scepticism of the status of the category 'nature' across cultures and certainly in the eastern Pacific. As Descola (1996) and others have noted, critically in the Oceanic context by Hviding (1996, 2003), the broader concept of nature has a spectacularly overdetermined and thorny history in Western intellectual traditions in which some conceptions of nature are highly privileged, both universalised and, to hazard a low-hanging pun, naturalised (Collingwood 1945). Harrison (1992) alludes to this conceptual tangle in the epigraph above, observing the dualistic Western approach to forest as nonpareil stand-in for the category of nature in Western thought, the fundamental ontology of the forest as other-than-human, while simultaneously capturing the critical anthropological position that ontological categories of natural kinds are socially constructed as well as culturally construed. This is no less true in the Pacific where the concept of nature has a rich treatment in regional studies. Consider Smith's (1985) classic unravelling of the representation and conventionalisation of nature and the natural in the age of early European Pacific encounters. Famously, he noted that early European encounters in the Pacific imposed the perspectival and representational frames of voyagers on islands while simultaneously contributing to the emergence of new Western romantic and scientific naturalisms (Smith 1985). Attention to Pacific literature suggests that individual authors, both scholarly and otherwise, have often bumped into the obscure and distorting character of putatively universal categories and characteristics of 'nature' because of the evident murkiness of their place or quality in historical encounters. Notably, such attention has not generally percolated down to more material scales nor yet penetrated academic works across disciplines often relying heavily on ontological concepts such as beach, bay, lagoon or forest as natural categories in their sense-making about Oceania (Hviding 2003: 249). What, then, must be done to bring insular ecosocial kinds into critical awareness?

Ordering the Gambier

> The kinds of forests dealt with ... belong to several different orders (Harrison 1992: 107).

While there are many distinct spaces of nature, and certainly a number of different understandings of and sentiments about nature in the Gambier, in company with the other works in this volume I focus on the ecosocial and ecocultural category of forest (*forêt*) or *tiarakau* because of the particular entanglement of imputable forested spaces in a range of critical collective issues in contemporary life. For instance, even something as fundamental as what counts as forest now has an active role in social processes in which Mangarevans construct and negotiate the making of selves and others. At the same time, the evident significance of the ambiguity of the underlying question of the object itself — what is forest? — reminds one of the continuing need for critical appraisals of 'nature' in the Pacific. Is forest a natural kind, that is to say, an ontological real? Or is forest an unnatural kind, yet another cultural imposition distorting anthropological attempts to make sense of Pacific lives and histories in terms of the material facts of the floral environment in the lives of the women and men for whom these islands exist as an immemorial home? Is there any such thing as forest in the Gambier, many copses and collectivities of trees notwithstanding? What might otherwise forest-seeming places be if not forest?

The concept of 'cultivated bay' is one starting point for considering the place of forests in anthropological sense-making in some islands of the eastern Pacific where upland slopes have sometimes figured (in European representations) as virtual deserts or, if perceived as forested, as undomesticated wildlands.[1] One notes the privileged position of this ecosocial category as a way to frame the intersection of culture and nature in Pacific archaeology, anthropology and history, and as the calibrating model of social order and island emplacedness. In the Mangarevan case, this is evident in Te Rangi Hiroa's (Sir Peter Henry Buck's) superlative *Ethnology of Mangareva*:

---

1   Alternatively, one might start with the concept of 'beach'. Indeed, just as forest is the nonpareil stand-in for 'nature' in Western ecosocial conception, so beach might be argued to be the nonpareil stand-in for island natures in much Pacific anthropology and history. That so much superlative work has been devoted to recognising and remaking the understanding of the beach as a site of social and cultural encounter from Malinowski (1984 [1922]) to Dening (1988) and beyond, as in Jolly et al. (2009), perhaps only underscores the need for further examination of its ontological status as a site of and in nature in its own terms, while highlighting the need for attention to other 'natural' spaces of the islands.

The *bay lands*, desirable as sites for dwellings and beaching places for canoes and rafts, with the introduction of cultivable food plants became extremely valuable, for it was only in such alluvial ground that the breadfruit, coconut, and banana could thrive. Such land was limited in extent and became vastly desirable. The desire for its possession created innumerable struggles and, apart from blood feuds, the dominant motive among chiefs and warriors was to obtain a share of cultivable land to give them a richer food supply. The introduction of cultivable food plants raised the standard of living for those fortunate enough . to acquire cultivable lands, while the wild plants of former days formed the complement (*kinaki*) to fish for the poor and the masses. The desire for better vegetable foods led fishermen to the economic pursuit of exchanging fish for preserved breadfruit. Cultivable foods, *kaikai 'akariki* (foods of kings or high chiefs), and wild foods, *kaikai a te oge* (foods of the hungry) became associated with the cleavage in social position between landowners or cultivators and the common people who were without cultivable land. (Buck 1938: 199)

On one hand, Buck offers nothing less than a penetrating anthropological model of the intersection of nature, culture and social order, well supported by the facts on the ground in numerous works from across Oceania. On the other hand, one might ask how does and does not 'cultured bay' map onto locally significant paradigms of human dwelling-in-place as the intersection of locally salient formulations of the spatial relation of nature and culture as *gnati* (bay, tribe), *nuku* (maternal lands) and *kaiga* (personal lands) in which the biosocial and the genealogical is mapped into the geophysical terrain that extends past the perceived-as-cultivated limits of bay spaces? How do such models as Buck's ignore or displace the margins of both cultivated and evidently non-cultivated spaces, boundaries between ecosocial zones to which a category of forest in eastern Polynesia may in some ways belong? Are they ethnographic objects to which sufficient attention has been paid in their own terms?

The area of cultivable land was small in comparison with the size of the islands, and lands and their produce were held by the ruling families and landowners who had been rewarded for services rendered in war. A large number of the common people having no cultivable land at all were thus denied direct access to vegetable foods, such edible wild plants as grew on the promontories and hillsides beyond them was the sea, which they tilled assiduously. A large portion of the commoners depended for the necessities of life on fish, shellfish, crustaceans, and such edible wild plants as grew on the promontories and hillsides beyond the boundaries of the cultivable lands. (Buck 1938: 197)

Again using Buck as a stalking horse — with no intention of suggesting his work as anything other than exemplary of a standard model of the nature–culture intersection from his anthropological era to the present — one might conclude that although wild land (including 'forest') is present in his model, its presence is largely context for a foundational position on human mastery and dominion over the landscape, centred on cultivated bays and ocean spaces in contrast, if not opposition, to imputed island-interior wildlands.

> The flora of Mangareva is disappointing. The steep slopes of the hills combined with the rocky soil are not conducive to luxurious growth. The scarcity of the flora is stressed by traditions that, when first peopled, the land was bare of trees from the beach to the foot of the hills, and that trees as well as cultivable food plants were introduced later by Tupa. (Buck 1938: 8)

While Buck addresses the situation and character of plant cultures early in his studies, his focal arguments rest on a native/introduced division of flora, and by the social relationships mapped onto distinctions in access to lands producing *kaikai 'akariki*, the chiefly foods cultivated in near-shore plantations owned by *togo'iti* elites in the pre-contact Gambier, as opposed to *kaikai a te oge*, six wild foods found on the upland and, from his point of view, non-domesticated spaces of these islands, normally only consumed by commoners.

Following the position established by Hau'ofa (1993) in other analytical domains, nature in the eastern Pacific perhaps seems meagre or too thoroughly domesticated and controlled, hence easily bracketed relative to continental chauvinism and the infelicitous baggage of Western intellectual traditions. For instance, that island spaces have diverse histories of use has been long evident in the anthropological literature. Yet a pervasive myth of the pristine hovers over many island spaces, a perverse legacy of the history of the Western romanticism of Pacific nature implying that domestication of the island ceased at some more or less arbitrary social boundary beyond which were wilds, and that many or all of these ecosocial spaces were fixed in extension or duration on the landscape.[2] In fecund reality, Pacific landscapes and ecologies were and are highly varied and may too often have been conceived in reductionist terms, as in Buck's categorisation of the Gambier's vegetable biofacts (Table 2.1).

The presence of a floral chauvinism evident in the history of the interpolation of the Gambier into Western scholarship via Buck's Mangarevan work is a case in point with his insistent focus on *economic flora* and human domestication. Notably, what makes some flora economic is nowhere defined. The utility of 'utility' as an analytic is taken as a given. Foremost, it offers Buck the means to undergird his interpolation of Mangarevan forms of rule and sociopolitical identity into a comparative anthropology, ultimately by grounding the distinction between common and uncommon classes in society in large part in terms of their relations to land and its produce. Landowning cultivators are *ragatira* or *togo'iti* (Buck 1938: 200), social elites who are authorised and

---

2   See also Hviding, Chapter 3, and Bell, Chapter 6, this volume.

legitimate possessors of useful, economic plant cultures largely imported during settlement periods and long domesticated. Contrastively, *'urumanu* (commoners) subsist on the fruits of the sea and what Buck labels *kaikai a te oge*, foods of the hungry, during lean times and good. Commoner engagements with, and use of, wildlands is relegated to an afterthought despite the evident significance of that part of the landscape in regularly supporting the many (Buck 1938: 215).

Table 2.1 Buck's categorisation of the Gambier Islands' vegetables.

Native plants	
'ara: *Pandanus* sp.	pokea: *Portulaca lutea*
gegie: *Suriana maritime*	potini: an edible fern?
kokini tekeo: *Tephrosia purpurea*	to'ira: *Cyperus pennatus*
koueriki: *Terminalia koariki*	
Plants introduced by Polynesian settlement	
'au: *Hibiscus tiliaceus*	miro: *Thespesia populea*
ere'i: *Cocos nucifera* (coconut)	nono: *Morinda citrifolia*
eute, pure: *Broussonetia papyrifera* (paper mulberry)	pia: *Tacca pinnatifida* (arrowroot)
rama: *Aleurites oluccana* (candlenut)	gatae: *Erythrina* sp.
rautea: *Ipomoea* sp.	kape: *Alocasia macrorrhizos* (taro)
rega: *Curcuma longa* (turmeric)	katiru: *Cucumis* sp.
tamanu: *Callophyllum inophyllum*	ke'ika: *Eugenia malaccensis* (mountain apple)
taro: *Colocasia esculenta* var. *antiquorum*	ko'e: *Schizostachym glaucifolium* (bamboo)
ti: *Cordyline terminalis*	to: *Saccharum officinarum* (sugar cane)
kumara: *Ipomoea batatas* (sweet potato)	tumei, mei: *Artocarpus incisa* (breadfruit)
'uatu: *Musa* sp. (plantain)	meika: *Musa* sp. (banana)
u'i: *Dioscorea bulbifera* (yam)	konini kai (edible creeper, not collected)
aoa: *Ficus umbilicata* (banyan)	kaka'o: *Miscanthus japonicus* (cane)

Source: Buck (1938: 9).

What might be thought of as forest as an ecosocial kind, in the normal sense of woodlands and wilds, a site for a certain kind of human encounter with nature in the Gambier, is thus in some sense defined out of existence despite the evident significance of wildlands to commoners (and *togo'iti* in lean times) after the early settlement period commencing with the arrival of the culture-hero Tupa, but before contact with European cultures. Lands are figured as either fully domesticated or fully wild (i.e., not economic, not playing a role in supporting social status). This is a characteristic dichotomy that I suggest deserves further examination across the eastern Pacific and Polynesia more broadly in classic works as well as more recent examinations of island ecologies.

Reordering the Gambier

The story of the Catholic instauration in the Gambier, a putative theocracy that re-architected these islands from the foundations up, has been told and retold since the mid-nineteenth century (Delbos 2002). The basic outline of the European interpolation of the Gambier into the nineteenth-century world system and Catholic development of Mangarevan society and lands, prefiguring effects on nature, can be quickly summarised. Within years of the 1834 arrival of evangelical and evidently extraordinarily charismatic priests, including the infamous *Péres* Laval and Caret along with an Irish lay brother, Catholicism was embraced by one of the main chiefly parties of the era, and used to solidify and consolidate the rule (*ao*) of the archipelago in a period of significant tensions between different chiefly families as well as within the royal line of the *'akariki nui*. This consolidation of power under the renewed paramount authority of the singular *'akariki* Te Maputeoa was spectacularly good for the mission and, if too transiently, for Te Maputeoa. The new order resulted in imputed improvements to the material infrastructure of the islands as well as the establishment and regulation of authority over the numerous traders, merchants, beachcombers and pearl exploiters, who had been arriving in ever-greater numbers, year after year, since the 1820s.

Practically, for Europeans arriving in the Gambier in the mid-1840s, the instauration then still in progress seemed a Pacific miracle, as if a provincial French village had been transported to the eastern Pacific, about as far from a continental landmass as is possible on this terraqueous globe. The impressive list of architectural feats achieved over mere decades by the population probably does little to convey the experience of then contemporary arrivals. The 2,000-seat cathedral in Rikitea, four other stone churches each capable of seating many hundreds, a two-storey, many-roomed college and seminary, the convent, prison, king's tower, chapels, stone-paved roads, quays, numerous follies including massive triumphal arches, and over 50 massive hand-quarried, faced-stone homes on a French provincial style, many of two storeys (Vallaux 1994).

The arrival of new food crops was, for Mangarevans at least, no less spectacular. Indeed, a key feature of the re-architecture of the islands was the establishment of new plantations, often in cultivation zones not previously used for what Europeans recognised as agriculture, the so-called wilds where *kai oge* or *ono oge* were harvested by *'urumanu*. Parts of the island not previously domesticated, as the term might have been conceived by Buck, were transformed or, perhaps, re-domesticated. Simultaneously, and perversely, parts of the islands previously highly domesticated, *gnati* sites of settlement, were de-domesticated, moved out of the category of *punui* village spaces into *vao* wilds or *mataeina'a*[3] (agricultural spaces) as the demographic tragedy of the decades between the 1820s and 1860s played out. These transformations contributed to, or played a role in, revaluations of ideas about domesticated and wild spaces in the archipelago.

With retrospective acuity, Delbos in his excellent *La Mission du Bout du Monde: La Fantastique Aventure des Bâtisseurs de Cathédrales dans l'Archipel des Gambier* (2002) argues that the Catholic spiritual instauration, the social and political reorganisation of the population, the implementation of the rule of law, and the stabilisation of the pearl trade were all a by-product of the architectural mission. It is as if the organisation of labour necessary to erect the edifice of the mission was self-fulfilling in realising the mission as a reordering of the social relations of the tribal groups along Western lines. *Ati*, as labouring units, no longer were to be conceived as established or grounded by their ritual and practical connection to specific *kaiga* and the breadfruit groves and storage pits ritually maintained therein, but by their orientation to and material maintenance of particular church structures in particular districts, including new plantations. Importantly, the massive cathedral was conceived and achieved as a collective labour by competitive work groups drawn from across the *ati*. The re-architecting of the legal realm operated to support and bolster the reorganisation of the social realm around labour rather than around kin relations within *ati* and shared connections to breadfruit groves and specific terrains, named and collectively utilised and literally incorporated into/as descent groups. Importantly, the reordering privileged more varied personal or individuated gardening around homesteads (*'are 'akairiga*) as well as greatly expanded cultivation.

The more the mission built, the more it required the fabric of Mangarevan culture, society and land to shape itself to the building of the mission, and the more the mission could build.[4] The more it cultivated, the more it could cultivate — culminating in the establishment of orange groves, a significant raspberry plantation and widespread coffee plantations, among other wildly new cultivars. Within years of its instauration, Mangareva was widely discussed in a number of maritime and Catholic and mission journals and papers, especially

---

3   Modern, borrowed from Tahitian.
4   There were many forms of spatial transformation parallel to the translation of previously *vao* spaces into newly 'domestic' plantations. For instance, the chief's dwelling was reconstructed as a king's tower and palace (conserving the structures of power within the same line of descent while transforming the tokens and conceptions of that power along European lines), the traditional circum-island *ara* was cobbled and converted into the *Ara Nui* (privileging island traffic along a *mua/roto* axis as never before) and part of the terrain of and below the royal nurseries *Te One* was converted into the convent *Rouru* (maintaining while transforming the gender and sanctity of the land).

after d'Urville's visit, and praised for the rapidity and totality of the mission's achievements — compare Goyau's (1928) summary of the vast and varied nineteenth-century commentary.

Delbos (2002) draws attention to the process by which Mangareva, a 'New Paraguay' as it was seen by visiting dignitaries beginning with d'Urville, Moerenhout and others not a decade after the commencement of the mission, was effected because of the curious dovetailing of the architectural and engineering skills, presumably coincident to their spiritual talents, possessed by several of the key figures of the early mission. The mechanics of the transition from a stone-age society to a model Catholic community, he argues, were concretely achieved as architectural project. Mangareva became, as he puts it, a land of *bâtisseurs* (builders). This may well have been the case, but I note it was also becoming a land of *jardiniers* (gardeners) — compare Laval (1968) for numerous anecdotes regarding the influence of introduced cultivars on the overall success of the mission. Thus the westernisation of the islands was also a re-domestication, a process of reordering nature in culture that was characteristically individuated (non-collectivist), democratised (the disestablishment of chiefly privileges over *kai 'akariki* alongside the vast expansion of available cultivars), and an expansion of domestic lands and subsequent if temporary diminishment of the wilds. Initially *vao* spaces became domestic plantations for new foodstuffs that greatly expanded and enhanced the kinds and numbers of domestic groves on the island even as traditional socially foundational near-shore breadfruit groves receded in significance and collective utility and were eventually abandoned.

But the autopoietic success of the mission as a total reinstitution of Gambier society was not without its dark irony. Even before the beginning of the instauration, and continuing throughout the entire middle of the nineteenth century, the Mangarevan population declined dramatically. By the time the last bricks were being installed at the tops of the cathedral and church towers, the keystones placed, the arches erected, the new plantations of coffee and raspberry and orange established, there were scarcely persons enough to fill one of the four churches, much less the cathedral (Vallaux 1994; Kirch and Rallu 2007). Meanwhile, commercial interests on the island, centred on the pearl and reprovisioning trades for passing navigation, began to chafe under the impositions of the Code Mangarévien — essentially a Western legal support to the total architectural institution of the mission which sought among other things to co-opt labour for the slightest material or spiritual transgression. Ultimately, and almost simultaneously, the mission project and the chiefly *ao* (rule) of the Gambier collapsed under the tensions of the demographic tragedy. New wilds grew up and subsumed all of these works, architectural or vegetable, engendering new forms of ecosocial space. Lands only just (re)domesticated over the previous decades now returned to, or became, wilds or *vao*. What might have seemed an expansion of wildlands appears a process of *de-domestication* of previously thoroughly human spaces.

Wildlands and Forests

> During the early middle ages the northern forests of Europe were vast, stretching across the continent like domes of darkness and the indifference of time. Interspersed throughout them were smaller or larger settlements lost in the shadows of antiquity's decline. With respect to the medieval social order that was reorganizing itself on the basis of new feudal and religious institutions, the forests were <u>foris</u>, 'outside'. (Harrison 1992: 61)

The extensive anthropological literature on culture and ecology in the Pacific has generated a number of enduringly potent arguments. Yet, a centring glimpse forward and back from exemplary mid-century texts such as Goldman (1955, 1957, 1958) or Sahlins (1958) suggests that external commentators have privileged the adaptive, ecologically canny, masterful character of the human–ecological interface on many remote Pacific islands with respect to the role of ecology in political and social organisation, particularly complexity and stratification. Professional and amateur observers alike have often been profoundly influenced by and made much of such claims along the lines of 'look what *they* have done with so little' from the age of early contacts with European voyagers, through the ecological turn in mid-century Pacific anthropology, and certainly such perspectives are shot through contemporary archaeological framings of the region. Moreover, such a perspective is often imbued in works, including by the same authors (cf. Kirch and Sahlins 1994), that considered the sociocultural and political transformations engendered by the sandalwood rush in the eastern and central Pacific in the early nineteenth century, the first post-contact era inter-island interpolation of Pacific forests into global trade, as exemplary of human dominion over island natures.

Certainly this position has its attractions. Who would not thrill to the story of the (pre/post) Lapita conquest of ocean and its lands and seaways (Kirch 1989, 2002)? But it may also be a remarkably enduring form of romanticism, a framing narrative that directs the kinds of questions we ask of others, of our field data, of our works. As has now been thoroughly discussed, Western scholars of others' cultures project our fundamental ontological categories of nature space at significant analytical risk (Descola 1996; Selin 2003; Lotz 2005; Escobar 2008). Broadly circulating invocations of nature as where man is not or where culture is not and of island spaces aside from 'cultured bays' as wildlands, abcultural or boundary-cultural spaces useful only for purposes of resource extraction of famine (undomesticated) foodstuffs and certain material resources, place anthropological analysis at particular risk.

In conversation with this literature — importantly Hviding (1996: 170), who notes that a wild/tame analytical pair can be enhanced by a sensitivity to degrees of domestication — this chapter has been drawing attention to the analytical utility of locally diverging notions of used and unused, inhabited and uninhabited, cultivated and uncultivated spaces, significant in the fashioning of place and person, as a potential amelioration of the ontological challenge of mediating 'nature' and 'forest' in anthropological sense-making. In the Gambier, the paired terms *punui* and *vao*, domestic space and wilds, offer a meaningful distinction that can perhaps serve to tease out a meaningful sense of forest among other contemporary *vao* spaces. The relevance of this pair is that it can be evaluated in terms of a domesticated–undomesticated distinction with an imputable pan- or meta-cultural interpretability. Acknowledging that the presence of humans on the landscape (or seascape) fundamentally disestablishes claims that nature, or forest for that matter, is where man is not (Vogel 1996; Hull 2006), I note that domestication offers a gradient, non-absolute, conceptual framework for making sense of nature's spaces in and across cultures. In essence a process, not a static state, as an analytical framework, domestication reflects the reality of biological and human cycles in time.

In the Gambier, contemporary Mangarevan understandings of the relation between highly charged social spaces and equally charged absocial island spaces are categorised in the *punui/vao* pair. The category of *punui* is relatively unproblematic: (1) a village, a town; the chief settlement or capital (Tregear 1899) or n. *ville, village; endroit principal d'un pays* (Rensch 1991). The word *vao*, however, is somewhat more nuanced: (1) uninhabited lands; lands unplanted and uncultivated, (2) a plain; open country without trees, (3) relationship; family (Tregear 1899).[5] Or, in Rensch's (1991) French/Mangarevan dictionary: *Vao* n. *champ, endroit; terre non habitée et non plantée; clan, tribu*. Intriguingly, *punui* does not appear to contain a semantic field relation to the concept of consubstantial sociality or group membership, while the category of *vao* does. In this sense the semantic pair for *vao* is *ati/gnati* — a Mangarevan synonym that means at once 'bay and people'.

Here is the rub. How can both domesticated space and undomesticated space overlap with semantic conceptualisations of a human group, an 'us'? How does this semantic inconsistency complicate such nature–society divisions as evident in Buck's model above? The possibility that *vao* society was constituted by *ati*

---

5   *Vao ke*, not to be of the same race or family (Tregear 1899).

disposed of desirable domesticated baylands and forced only to live on *kai oge*, meagre wild foods, is specifically excluded in all ethnographic accounts of the archipelago, including those by Buck. The answer, I learned from a number of senior Mangarevans, is still very much in memory and, somewhat, in practice. Traditionally, *vao* would not be used to refer to one's own group or to the groups of others in polite conversation. *Vao* space is 'other space', both in the sense of Buck's putatively undomesticated wild and in the sense of other people's lands to whose domestic fruits one is not entitled. Wildlands call into mind, from the perspective of some discursive context in which questions of personhood and collective belonging is at question, wild others. One's own group is always and only one's *ati* (*gnati*).

That Mangarevans' notions of domesticated/undomesticated lands were also mapped onto differences between human groups is highly consistent with Hviding's argument that a lack of cognitive dichotomy between the human/non-human characterises Oceanic views of 'nature' (2003: 259) and is ethnographically comparable to the Morovo case, in other ways culturally quite distinct, among whom the notion of *puava* (territory) 'does not constitute a distinct realm of "nature"' (1996: 170) vis-à-vis notions of collective identity, group membership or being *butubutu* (gens).

When I asked a group of friends in the summer of 2009, senior fixtures in Mangarevan society, *e aha te makararaga o te 'u teito o te vao*? — what are the thoughts of the ancestors about the *vao*? — two very different comments emerged over the course of an interesting conversation. First, as one man put it, *a tau ara* (at that time) Mangarevans 'lived so close to nature, there was no difference'.[6] He explained that a group somehow equates to its *vao*, its place in nature; a sense that the people of the past were in such a relation to their lands as to be categorically themselves others, in effect not his people. This is an extraordinary claim, I believe, one that is highly revelatory about his sense of contemporary practices and relations to the landscape as alienated from some imagined more authentic connection with lands now somehow displaced. Other participants pointed to another sense. As another man put it, 'crossing lines is dangerous', *e tinai ona*; one can be struck and killed, further highlighting the implied distance — in time, space, social relation — between the honoured *tupuna* (ancestors) and their contemporary inheritors. *Vao* spaces were a metaphorical line between the culture of one's own people and the dangerous other, whether human or natural or divine.

---

6   Note that my informant used 'they', not 'we', as he often did when speaking of *tupuna* ancestors.

Displaced Natures

> If it is true that the post-Christian era detaches itself from the past, frees itself to some extent from the inertia of tradition, 'comes of age' under the auspices of reason, it is also true that it experiences its freedom as a deprivation as well as a gain. Early in the last chapter we saw how freedom from the past implied freedom for an enlightened future. The countercurrent of Enlightenment's drive to inherit the future is nostalgia. As the ancestors fall silent in their graves; as the age-old traditions and landscapes of the past recede into vanishing horizons; and as the sense of historical detachment begins to doubts its original optimism — nostalgia becomes an irrevocable emotion of the post-Christian era. (Harrison 1992: 155)

Long before the mantle of the nuclear age was draped — in French red, white and blue — over the Gambier in the 1960s, with various sorts of fallout once again altering the local understanding of social–human spaces and wild–natural–dangerous spaces in the Gambier, these islands were already sacrificed with other east Polynesian archipelagoes on the altar of history. Ship-to-shore vectored diseases, the transformation of social organisation during the Catholic instauration and increased mortality from commercial pearl diving all bespeak a nearly unimaginable human tragedy. But even in the absence of a concrete accounting of demographic casualty, we can be quite certain that the population saw a 40-year decline beginning in the mid-1820s, when a difficult to estimate population would have been reduced by more than 60 per cent and perhaps greater than 90 per cent (Kirch 2007; Kirch and Rallu 2007). Careful records of Mangarevan births and deaths were instituted in the heyday of the mission and record the population decline in great detail until the 1870s, after which the population largely stabilised. For instance, a population of approximately 1,500 in 1863 had become 446 by 1881, and one notes there had already been several major epidemics before 1863, including one prior to 1834 which may have killed half the population at that time (Laval 1968). Also, it is not clear in the missionary census records — the chief source of pre-1881 demographic information — whether Rapa Nui and other Polynesians who at various post-contact periods constituted a significant portion of Gambier society were ever enumerated, although European, Chinese, Demi and non-Polynesians generally were.[7]

By the 1860s, about 500 persons still lived in the archipelago, some of whom had started life elsewhere in the eastern Pacific. Notably, this remained the stable population point until the 1980s, when the total number of inhabitants began to increase. At times, chatting with friends and other informants, it

---

7   Thus, while empirical, the question of whether the numbers in the early records indicate the actual Mangarevan population or all 'native' Polynesians remains intractable. If the recorded population figure includes all Polynesians living on the island, the specifically Mangarevan population must have been very small indeed.

seemed my fieldwork in a period of rapid demographic expansion was a post-apocalyptic anthropology, as the visual, auditory and even tactile trace of a past that had somehow not become the present regularly confronted me, so many *memento mori* in the words of informants, particularly when working with folks to clear a plot of forest for a house site, garden, old roadway, lands around one of the useless beautiful stone churches, or for aesthetic purposes. One might well say that contemporary Mangarevans, like their cousins on Rapa Nui and in the Marquesas, are uncertain descendants of survivors of the end of a world and at times in everyday conversations they seem conscious of it, and no more than when confronted in the bushy landscape by the architectural ruins of the Catholic mission alongside even more ancient stone ruins of the pre-Christian era(s).

Indeed, by the time the French nuclear test establishment, the Centre Experimental du Pacifique, established a large footprint in the Gambier in the mid-1960s, identified by Mangarevans as a critical watershed in their relationships to their island's ecospaces, the hand-cut stone pavements of the cross-island roads were showing immense neglect, few families lived outside the two *punui* previously constituted as competing high chiefdoms but consolidated after European arrival, and many people were already living exclusively in Rikitea. The watershed was much earlier, however. At some point after 1920, the traditional corporate descent groups called *ati*, which served as the primary frame of social organisation, finally dissolved into a small number of closely related families — today, a proverbial witticism suggests there may be only three. By the 1920s, dusk had settled on the marvels of the Catholic instauration that had earned the Gambier international fame in the mid-nineteenth century, the abbey finally emptied, the college abandoned, and the draughty and ageing stone houses left for newer 'plantation' style wooden houses. These houses were in turn abandoned in the last decades of the twentieth century for more contemporary concrete structures, although two or three of the dilapidated wood structures can still be seen, including the mothballed childhood home of the French Polynesian president Gaston Flosse.

One result has been that Mangarevan understandings of the relationship of non-domesticated space and lands, *te vao*, to domestic spaces and human places, *te punui*, have changed. For much of the past two centuries, many of this archipelago's spaces have increasingly been perceived as '*brousse*', 'scrub', 'bush' or 'forest'. This process peaked when the establishment of the French nuclear test regime in the nearby atolls Mururoa[8] and Fangataufa in the 1960s saw the end of any remaining intensive plantation agriculture along the de-peopled coasts. Concurrently, the population was reconcentrated in a single village adjacent

---

8   Also, sometimes, Moruroa Atoll.

to a hastily erected nuclear fallout shelter after fallout events in 1966, results of France's Pacific nuclear establishments (CESCEN 2006). At the same time, many of the previously densely inhabited and cultivated bays, and the slopes upland from them on Mangareva and across the archipelago, were subjected to an intensive and dramatically efficacious forestation project promoted and partially implemented by the regional administration's *Service de l'Agriculture* (Vallaux 1994: 127). In less than four decades, as visible in Figures 2.2 to 2.6, the long unfolding de-domestication of a substantial portion of the archipelago was finally consummated by a forestry project that seeded many of the upper slopes of the islands with successfully invasive pines, and sometimes more colourful *Albizia lebbeck* and *Albizia falcata*, greatly decreasing the ease of travel between bays or across the spine of the mountainous ridgeline and literally obscuring the face of the land.

Figure 2.2 Scene of near-shore cultivated bay lands on Mangareva at the time of d'Urville's visit to the Gambier, by Louis Le Breton, 1842. The rich, readily recognisable near-shore domestic plantations contrast with the dry, low-coverage upland slopes on all the higher islands of the Gambier at that time.

Source: Plate no. 42 in the Atlas Pittoresque of J. Dumont d'Urville's *Voyage au pole sud et dans l'Oceanie sur les corvettes l'Astrolabe et la Zelee.*

Figure 2.3 Mount Duff, Mangareva Island, early twentieth century.
The village Rikitea in the great bay is just out of site on the right. One notes
the near-absence of trees, commented upon by both Buck (1938) and his
Bishop Museum colleague Kenneth Emory, and an important context to
Buck's discussion of Mangareva's ecology.

Source: Photo by Te Rangi Hiroa (Sir Peter Henry Buck). From New Zealand Electronic Text Collection
under a Creative Commons licence, nzetc.victoria.ac.nz/tm/scholarly/BucViki-fig-BucViki_P017a.html.

Figure 2.4 Mount Duff with Rikitea village and bay in foreground, just after
mid-century yet prior to the massive forestry work of the territorial Service
de l'Agriculture.

Source: Photo by Roger Curtis Green, 1960. From DigitalNZ archive under a share licence, www.digitalnz.org/
records/32431161?search[page]=3&search[text]=mangareva.

Figure 2.5 While near-shore cultivated gardens and bays retain a traditional verdure, Mount Duff and downslope plateaus are now heavily forested, largely as a result of the regional agricultural forestry service's efforts.

Source: Photo by Alexander Mawyer, 2002.

Figure 2.6 While the vertical face of Taraetukura (Mount Duff, Figure 2.5) evidences a significant shift towards coverage, the island's spine and downslope ridgeline from Mount Duff's peak, running north-east, are now remarkably forested and the change is striking relative to any point earlier in the twentieth century.

Source: Photo by Alexander Mawyer, 2003.

Now the Gambier's population is burgeoning — 1,337 as of 2007 (ISPF 2011) — and families are beginning to push back into the de-domesticated (previously cultivated) bays, producing all sorts of curious entanglements with history, since the contemporary 'bush' turns out previously to have been someone's house platform, imputed *marae* or *papa* sitting stone or backyard. Sometimes it turns out that the bush is covering up one or more charming French provincial quarried stone houses from the height of the Catholic instauration between 1834 and the 1870s, long abandoned and generally forgotten, leading the contemporary Mangarevan landowners to ponder questions of historic loss over several different periods — not only the loss of tradition but of modernity around 1850, which somehow did not last and was absorbed into *vao* space. Such layerings of historical complexes are not restricted to the architectural. The old coffee plantations and the most incredible raspberry plantation on the back of Mount Duff similarly reverted to a historically murky bush quality, as did the remaining orange trees isolated across the archipelago and closely observed in their use-rights. Together, these all reveal a quality of Tsingian weediness (Tsing 2005)[9] in which these wooded terrains have become not only physically bushy, but also conceptually murky, socially fraught and affectively tangled.

In consequence of these histories, the ecosocial setting of the Gambier Islands, the relations between the places of the past and wild spaces, nature spaces, forest and bush, is not now easily disentangled. This tangled, weedy quality of the local understandings of the normative relations between Mangarevans and their islands' non-domesticated spaces became more visible in the years between 1999 and 2009, which were remarkable for the emergence of a new wave of highly domestic labour organised at the level of the *'akairiga* (the household family), which began during these years to take back the *vao*, the de-domesticated spaces of the islands abandoned over the past 150 years and notable for the overgrown ruins of the Catholic instauration and the all but impassable tropical gnarl that surrounded them. Land-clearing work parties in which I participated on four islands — with individual families, *'akairiga*, within groups of families claiming descent from the same *ati*, and with church groups representing the whole community — in 1999, 2002–03 and 2009 may have differed in the kinds of terrain from shore plots, to bay interiors and upland plateaus, and different historical distances from last moment of former habitation, but all· shared a characteristic sort of conversation. Discussions of these de-built and de-cultivated spaces frequently and tangibly produce anxiety in many speakers at many ages of life, centred on the question of what the facts of de-domestication and wilding of previously domesticated lands imply for their own Mangarevan

---

9   'Weeds have been of little interest to conservationists; we think of them only as indicators of disturbance' (Tsing 2005: 174).

identity and what might be called the descent-of-belonging or the inheritance of us-ness in conjunction with the, now generally, legal establishment of title to lands and the use of their resources.

Thus many folk in the Gambier today, old and young, have a complex relationship with the different sorts of island spaces common sense tells me to see as now-established forest with which they must engage in order to craft new homes on old lands. The issue, as differently positioned persons seemed to experience it, was thus the way that de-domestication seemed to erase or, at least, damage the descent through time of linkages to the past and its peoples — particularly their quality of true or authentic *tika*, Mangarevan-ness, with the unsettling implication that contemporary persons are *vao ke*, that is, social others. Again and again I have heard such questions posed: What should be understood of the decaying coffee plantations running up the ridges in the backs of the inhabited valleys of the main towns, the *punui*? What of the former church raspberry plantation, today difficult to reach, and free to all for the price of the effort to reach it, on the back of the island, *i tua*? What of the remaining orange trees, signs of futurity past here and there, each priceless? The 'ownership' of coconut groves, no longer cultivated? The few, modestly economised lychee? The actual *vao rakau*, wild, tangled, imposing? What is the place or role of the wild spaces in everyday life in the local conceptualisation of being-an-us, being Mangarevan? To whom do useful trees (coffee and orange, breadfruit, *hau* and *miro*) belong? To whom belong the forest-enclosed cultivars such as raspberry, signs of past domestications, from an era of collective possession and collective use? To whom belong those material things, legacies of past domestications, by forest obscured?

As in Harrison's epigraph at the start of this section, there was a historical collapse in the Gambier. But not of traditional Polynesian culture or society as Diamond (2005) would have it, and certainly not of Mangarevan culture of the *longue durée*. Rather, what collapsed was the post-contact colonial–theocratic order(ing) of things. Stone houses collapsed into ruins. Ancestors became others, plantations became plants, the *vao* became wilderness.

Other People's *Kaiga*

> It is hard for us to imagine, several millennia later, how the return of the forests was experienced as a cataclysm by many of our stone-age ancestors [after the last ice age] (Harrison 1992: 197).

Everyday life in the Gambier rests on a fabric of labour in the various sites, commercial and domestic, sacred and civic, terrestrial and maritime, that make up the manifest of human practices in these islands — this is the kernel of the long-enduring sense of domestication in its anthropological interpolation, that the human is a technologist, a reworker and remaker of an unruly nature, a pacifier of the world as given. We dwell in worlds of our own modification as *Homo domesticans*. Sociality can be loosely characterised as a function of work layered over a patina of private familial activities and occasional, generally predictable, civic activities annually centred around the summer cultural festival *Te Tiurai*.[10] So much for the thrum and throb of everyday life, social relations coalescing out of institutions of encounter and engagement in these various sites. Underneath all of this activity, Mandeville's buzzing hive, rests kinship, land and the social relationships affirming or denying shared rights to its use, particularly the use of its plant resources, the literal fruits of the land.

That land is politics is a banality that hardly requires comment. Since the unfolding of the rule of Western law beginning during the nineteenth-century administration of the Mangarevan kingdom (Code Mangarévien 1870), French land law was increasingly imposed on the Gambier. But the legacy of pre-contact and nineteenth-century land tenures, informal understandings between individuals and the church about rights to lands and their products, and the changing biofacts of the island's 'natural' spaces have resulted in a murky situation bearing on the practical dimension of the domestic–undomesticated distinction, the use of lands and the vegetable resources found on them.[11] The slippage between these two land tenure systems offers another perspective on undomesticated space. Nature's spaces, the contemporary wilds, are as contested now as they were in the centuries before contact with Europeans vastly expanded the variety and kinds of domestic foodstuffs available for cultivation in the Gambier, while coincidentally leading to the depeopling, deculturation and de-domestication of many of the archipelago's bays. In contemporary moments of everyday heightened significance, land disputes and land clearing, and construction and development, lead to personal and often intimate reflections on Mangarevan society, its generational and genealogical cohorts, and the meaning of the past in the present. The fallout is that kith and kin, to say nothing of neighbours almost inevitably more or less related (contemporary proximity predictably indicates past, if not present, kinship affinity), are

---

10   *Te Tiurai* (July) is preferred here to *Te Heiva*, which has specifically Tahitian overtones, locally problematic in the politics of inter-island personhood.

11   In principle, today a finite number of regional and national statutes govern land ownership and rights of use, property and propriety with respect to the lands of others. In practice, the situation is complicated and can be confounding. For instance, among other factors, the cultural order bearing on names of lands and names of persons may be more important to individuals thinking about the meaning of land and their rights to the products of the land. As a judge on Mangareva in 2002 as part of a multiyear process of settling title on Gambier lands affirmed, it is by no means certain that even the latest statutes which in theory have modernised the registration of property in French Polynesia will be resolvable by the courts within the time frame of five to ten years as anticipated by the Territorial Assembly. 'Perhaps', he wondered, 'thirty years will not be enough.' Until then, land tenure and the politics of the everyday that revolve around it will remain, he suggested, entangled in the legacy of multiple understandings, codes and practices.

caught in a web of increasingly fraught relations as shared connections to *nuku*, 'maternal lands', are transformed, as the land itself is being transformed by increasingly successful legal regimes into *kaiga* — mere 'land' in a legal sense. Such transformation further heightens for contemporary cultivators and domesticators their sense of a disconcertingly uncertain relation to legitimate inheritance or descent.

Conclusion

In response to the concept of forest in Oceania, this chapter suggests that at least part of the significance of this category can be its role in contributing to a critical reappraisal of nature spaces at the intersection of local and extra-local understandings of self and other, past and present, in the contemporary Pacific. As Hull (2006) notes in his intriguing *Infinite Nature*, there are myriad ways that people can relate to nature. The category of forest, *forêt*, *tiarakau*, is but one, as is *vao*, and attention to such categories does more than articulate how it is that nature is socially constructed, but offers a lens onto human practical relationships (themselves domesticated and undomesticated) with their lands and personhood.

Aside from its role in framing a dark romanticism in European literature and art (à la Melville or Gauguin), the forest in eastern Polynesia has been something of a categorical nullity. I suggest that this is surprising given the established importance of the early nineteenth-century sandalwood trade on some of this region's islands in establishing early structural conditions upon which many later social and political developments rested. Meanwhile, across archipelagoes of this region such as the Gambier, the critical ecosocial role played by breadfruit groves and the surrounding highly domesticated seaward lands prior to contact with European cultures appear rarely to have been perceived or anthropologically interpolated as forested in favour of conceptions of domesticated plantations for 'economic' trees and wildlands for all other ecosocial spaces in the islands. Uplands were seen as more or less undomesticated wilds of last resort, non-economic, regardless of whether they were wooded or not.

This chapter suggests that the categories undergirding the process of sense-making of nature spaces in these island contexts (both local and extra-local) have gotten bushy, or weedy in one of Tsing's (2005) senses. The changing biofacts of nature in the Gambier Islands have not merely resulted in the obscuring of the land under a patchwork of verdant weediness due to depopulation, de-domestication and late-twentieth-century new forestry projects, they have also resulted in an obscuring of locally salient intimate understandings of selves and others. The character of relatively non-domesticated forest as natural in the Western conceptual sense as a place where man is not is ultimately suspect in the Gambier, as it is perhaps elsewhere in the Pacific (if not everywhere humans have dwelt over the tumbling millennia). Significantly, local cultural understandings of nature spaces draw meaningful distinctions between kinds of domains with significant implications for human sentiment, practice and experience along domestic–wild categories.

That the concept of forest poses an ontological question in the Pacific is, as I think all of the pieces in this volume demonstrate, well established. That the fact of the forest poses a practical problem, or a matrix of issues, challenges and contestations, also seems evident. Like many readers across disciplines, I have been influenced by Harrison's (1992) superb reflection on the unfolding character of the forest for actual human populations and experience in European history, both for its particularity and applicability to the encounter with the *foris*, the outside, the verdant wilds, elsewhere and elsewhen. I argue that a central, and as elsewhere in the eastern Pacific too little discussed, aspect of the experience of the Gambier Islands since cultural contact with Europeans from the early nineteenth century has been a transformative process of de- and re-domestication. I propose that attention to the cyclical, dynamic character of this model of Pacific nature, including Pacific forests, can contribute to the disestablishment of too-narrow linear narratives of island development both before and after contact. Since that period, the process of the de- and re-domestication of the Gambier has been experienced in dimensions eminently symbolic, requiring conceptual adjustments as well as pragmatic responses.

The use of fruiting trees, bushes and vines, the sense of their place in the landscape, or their present use and rights thereto, even where they are and what they mean, is today erupting into everyday conversations and practices and informing varying intense encounters in the micropolitics of intra-island society. Discussion of the meaning of such de-built and de-cultivated spaces hinge everyday conversations to politically heightened issues of civility, sociality and Mangarevan identity. The Mangarevan word for forest, *tiarakau*, points more or less to land planted with trees, with an emphasis on agency, utilisation, use-value and other degrees of domestication somewhat at odds with core features of the category 'forest' in Western and, I think, much anthropological conception. In this chapter, I have drawn attention to the social significance of what otherwise might appear to be forest in the Gambier as *vao*, an outside that is in structural tension with the inside, the Mangarevan 'we', both practically as a matter of land-use and property rights, and conceptually in the way in which a sense of self undergirds collective existence put at stake. In this part of the Pacific, as perhaps elsewhere in the region, the enduring absence of a categorical or conceptual dichotomy between lands and persons, as metonymical between nature and culture, yields a characteristic form to this problem. What is the *vao* now? Are our spaces our own? Are we who we thought we are when our woods are wildlands?

References

Anderson, A., E. Conte, P. Kirch and M. Weisler, 2003. 'Cultural Chronology in Mangareva (Gambier Islands), French Polynesia: Evidence from Recent Radiocarbon Dating.' *Journal of the Polynesian Society* 112(2): 119–40.

Buck, P.H., 1938. *Ethnology of Mangareva.* Honolulu: Bernice P. Bishop Museum (Bulletin 157).

CESCEN, 2006. 'Rapport de la Commission d'Enquête sur les Conséquences des Essais Nucléaires en Polynésie Française par l'Assemblée Territoriale de la Polynésie Française.'

Code Mangarévien, 1870. Manuscript.

Collingwood, R.G., 1945. *The Idea of Nature.* Oxford: Oxford University Press.

Delbos, J.-P., 2002. *La Mission du Bout du Monde: La Fantastique Aventure des Bâtisseurs de Cathédrales dans l'Archipel des Gambier.* Papeete: Éd. de Tahiti.

Dening, G., 1988. *Islands and Beaches: Discourse on a Silent Land: Marquesas, 1774–1880.* Chicago: Dorsey Press.

Descola, P., 1996. 'Constructing Natures: Symbolic Ecology and Social Practice.' In P. Descola and G. Pálsson (eds), *Nature and Society: Anthropological Perspectives.* New York: Routledge.

Diamond, J., 2005. *Collapse: How Societies Choose to Fail or Succeed.* New York: Penguin.

Douglas, B., 2009. 'In the Event: Indigenous Countersigns and the Ethnohistory of Voyaging.' In M. Jolly, S. Tcherkézoff and D. Tryon (eds), *Oceanic Encounters: Exchange, Desire, Violence.* Canberra: ANU E Press.

Escobar, A., 2008. *Territories of Difference: Place, Movements, Life, Redes.* Durham (NC): Duke University Press.

Goldman, I., 1955. 'Status Rivalry and Cultural Evolution in Polynesia.' *American Anthropologist* 57(3): 473–502.

Goldman, I., 1957. 'Cultural Evolution in Polynesia: A Reply to Criticism.' *The Journal of the Polynesian Society* 66(2): 156–64.

Goldman, I., 1958. 'Social Stratification and Cultural Evolution in Polynesia.' *Ethnohistory* 5(3): 242–9.

Goyau, L.G., 1928. *Le Premier Demi-siècle de l'Apostolat des Picpuciens aux Iles Gambier*. Paris: Gabriel Beauchesne.

Harrison, R.P., 1992. *Forests: The Shadow of Civilization*. Chicago: University of Chicago Press.

Hau'ofa, E., 1993. 'Our Sea of Islands.' In E. Waddell, V. Naidu and E. Hau'ofa (eds), *A New Oceania: Rediscovering Our Sea of Islands*. Suva: University of the South Pacific, School of Social and Economic Development.

Hull, B., 2006. *Infinite Nature*. Chicago: University of Chicago Press.

Hviding, E., 1996. 'Nature, Culture, Magic, Science: On Meta-Languages for Comparison in Cultural Ecology.' In P. Descola and G. Pálsson (eds), *Nature and Society: Anthropological Perspectives*. New York: Routledge.

Hviding, E., 2003. 'Both Sides of the Beach: Knowledges of Nature in Oceania.' In H. Selin (ed.), *Nature across Cultures: Views of Nature and the Environment in Non-Western Cultures*. Dordrecht: Kluwer Academic Publishers.

ISPF (Institut de la Statistique de la Polynésie Française), 2011. www.ispf.pf.

Jolly, M., 2009. 'The Sediment of Voyages: Re-Membering Quirós, Bougainville and Cook in Vanuatu.' In M. Jolly, S. Tcherkézoff and D. Tryon (eds), *Oceanic Encounters: Exchange, Desire, Violence*. Canberra: ANU E Press.

Jolly, M., S. Tcherkézoff and D. Tryon (eds), 2009. *Oceanic Encounters: Exchange, Desire, Violence*. Canberra: ANU E Press.

Kirch P., 1989. *The Evolution of the Polynesian Chiefdoms*. Cambridge: Cambridge University Press (New Studies in Archaeology).

Kirch, P., 2002. 'Three Islands and an Archipelago: Reciprocal Interactions between Humans and Island Ecosystems in Polynesia.' *Earth and Environmental Science Transactions of the Royal Society of Edinburgh* 98(1): 85–99.

Kirch, P., 2007. *On the Road of the Winds: An Archaeological History of the Pacific Islands before European Contact*. Berkeley: University of California Press.

Kirch, P.V. and J.-L. Rallu (eds), 2007. *The Growth and Collapse of Pacific Island Societies: Archaeological and Demographic Perspectives*. Honolulu: University of Hawai'i Press.

Kirch, P. and M. Sahlins, 1994. *Anahulu: The Anthropology of History in the Kingdom of Hawaii, Volume 1: Historical Ethnography*. Chicago: University of Chicago Press.

Laval, H., 1938. *Mangareva, l'Histoire Ancienne d'un Peuple Polynésien*. Paris: Librairie Orientale Paul Geuther.

Laval, H., 1968. *Mémoires pour Servir à l'Histoire de Mangareva, Ère Chrétienne 1834–1871*. Paris: Musée de l'Homme (Publications de la Société des Océanistes 15).

Lotz, C., 2005. 'From Nature to Culture? Diogenes and Philosophical Anthropology.' *Human Studies* 28(1): 41–56.

Malinowski, B., 1984 [1922]. *Argonauts of the Western Pacific*. Long Grove: Waveland Press.

Rensch, K., 1991. *Tikianario Arani-Mangareva/Dictionaire Mangarevein-Française*. Canberra: Archipelago Press.

Sahlins, M., 1958. *Social Stratification in Polynesia*. Seattle: University of Washington Press (Monographs of the American Ethnological Society 29).

Selin, H. (ed.), 2003. *Nature across Cultures: Views of Nature and the Environment in Non-Western Cultures*. Dordrecht: Kluwer Academic Publishers.

Smith, B., 1985. *European Vision and the South Pacific*. New Haven (CT): Yale University Press.

Tregear, E. 1899. *A Dictionary of Mangareva (or Gambier Islands)*. Wellington: John Mackay, Government Printing Office.

Tsing, A., 2005. *Friction: An Ethnography of Global Connection*. Princeton (NJ): Princeton University Press.

Vallaux, F., 1994. *Mangareva et les Gambier*. Tahiti: Etablissement Territorial d'Achats Groupes.

Vogel, S., 1996. *Against Nature: The Concept of Nature in Critical Theory*. Albany (NY): SUNY Press.

3. Non-Pristine Forests: A Long-Term History of Land Transformation in the Western Solomons[1]

EDVARD HVIDING

Forests of New Georgia: Understanding the Long Run

The rainforests of the large mountainous islands of Melanesia are routinely assumed by outside observers to be an example of last remaining wilderness, high in biodiversity, valuable to humankind in general, and somehow largely undisturbed by humans until logging operations escalated in recent decades. But the fact that most Melanesian islands have substantial numbers of people living inland, or at least had inland populations until quite recently, would undermine such assumptions. As Bennett (2000: 18–26) expresses in her

---

1   This chapter builds on a larger multidisciplinary study of the long-term dynamics of human interactions with the rainforest in the Marovo Lagoon area of the Western Solomons, carried out with Tim Bayliss-Smith and the late T.C. Whitmore from 1996 onwards (Hviding and Bayliss-Smith 2000; Bayliss-Smith et al. 2003; Bayliss-Smith and Hviding 2012, 2014a, 2014b). Details of this study will be given below. As anthropologist, geographer and forest botanist, the three of us were able to share our field observations and insights from the Solomon Islands 'logging boom' of the 1990s, as well as to relate our longer-term experiences from the scenes of forestry and land use in Solomon Islands, in Whitmore's case reaching as far back as to his work as a government forest botanist in the British Solomon Islands Protectorate in the 1960s (Whitmore 1969). I am grateful to Tim and Tim for all those inspiring discussions in Cambridge and elsewhere. In this context I also would like to express my deep respect for the late T.C. Whitmore's continuous insistence as a botanist on the human factor in rainforest dynamics.

history of human interactions with forests in Solomon Islands, the rainforests encountered by the first settlers of the archipelago were 'opened up' by them tens of thousands of years ago: 'To the ancestral Melanesians and their more recent counterparts, the forests were a huge garden, with some part, for a time, being more intensively used than others' (ibid.: 19). From the 'arboriculture' and modest modifications by early human settlers, through the development of shifting agriculture based on burned clearings to present-day large-scale land transformations, the forests of the Solomons (and elsewhere in Melanesia and Oceania more widely) have histories that are closely interconnected with human habitation and activity in those forests (e.g., Bayliss-Smith et al. 2003). The continuous human agency in the long run of forest history has interacted ecologically with the fluidity and dynamic characteristic of the tropical forest itself, including its ability to regenerate after significant disturbance (cf. Whitmore 1990).

The forests of the New Georgia group of the Western Solomons, which are the focus of this chapter (Figure 3.1), have remained of particular interest both to logging operators and conservationist organisations, while consistently providing the majority of timber exports from the country since the 1980s. In New Georgia, the Asian (and increasingly Melanesian) protagonists of the Solomon Islands 'logging boom' that accelerated in the 1990s (Bennett 2000; Hviding and Bayliss-Smith 2000), as well as their colonial British precursors, saw islands with scattered seashore habitation, modest areas of forest clearance for shifting cultivation on the coastal slopes, and no inland villages. Dense rainforest still extended from coast to coast across steep ridges, extinct volcano craters, limestone pinnacles and mountain peaks, some reaching 800–1,000 metres above sea level. This, obviously, was also the deceptive scenery encountered by the conservationist adversaries of the loggers.

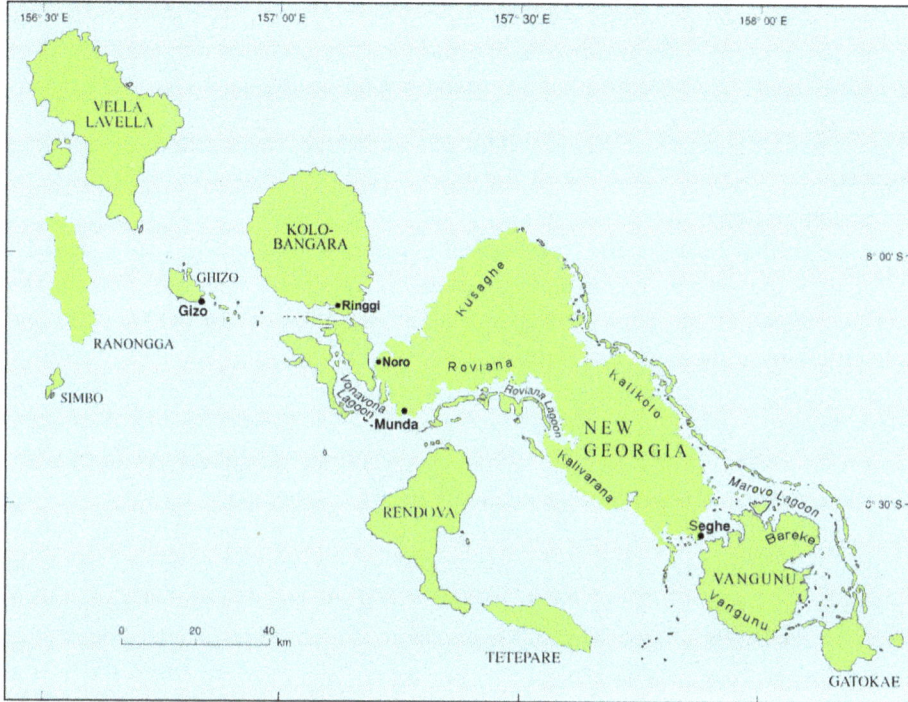

Figure 3.1 The New Georgia Islands, Solomon Islands.
Source: K.H. Sjøstrøm, University of Bergen.

What was not evident to outside observers was that the forests of New Georgia were in no pristine natural state when large-scale logging commenced. Despite the apparent intact quality in the 1980s of the New Georgian rainforest canopy, the forests were ecosystems in which the interaction with, and transformational influence of, human populations had played a major role through thousands of years. In fact, until the end of the nineteenth century the vast inland tracts of rainforest that were to become the subject of so much contestation were lands under extensive shifting cultivation and intensive river-fed irrigation, supporting large populations of inland dwellers. The assumedly undisturbed natural forests of late twentieth-century New Georgia had therefore regenerated from widespread, continuous human disturbance that only abated in the late nineteenth century (Bayliss-Smith et al. 2003). In this observation from the Western Solomons lies a challenging but potentially fruitful analytical interaction in the study of long-term land use, between separate paradigms of political ecology and historical ecology (see Blaikie and Brookfield 1987; Balée 1998; Robbins 2004). In the case of the New Georgia Islands, with obvious bearings for much of the Melanesian region, the understanding of the complex present-era politics of resource use and related conflict through 'political ecology' needs to rely more strongly on the truly long-term perspectives of 'historical ecology' concerning human–environmental interactions. Insights from botany are also a required element in the deepened understanding of rainforests as non-pristine.

Let me shift to recent political issues for the Western Solomons, New Georgia and the Marovo Lagoon. These are well-known fields of contestation that represent a common pattern in contemporary Melanesia. It is in fact not only the logging companies that have hit the forests of the Western Solomons in recent decades. While in the 1990s, Asian companies made rapid inroads into forests that seemed untouched except for tracts of land that had already been logged over to some degree by a colonial company called Levers Pacific Timbers, a plethora of international conservation organisations also descended on the islands and their forests: particular attention was given to the famous Marovo Lagoon in south-east New Georgia, a lagoon that covers about 700 square kilometres and is fringed by a unique double chain of raised barrier reefs and by the high volcanic islands of the eastern half of the New Georgia group (Hviding 1996a). The collective agenda of these organisations, which among them had the major international players WWF (World Wide Fund for Nature) and Greenpeace as well as smaller non-governmental organisations that were in fact more or less extended arms of Australian and New Zealand government interests, was to present to the world the idea that Solomon Islands in general, and the lands surrounding the Marovo Lagoon in particular, were places of pristine, some would say virgin, rainforest that had never been significantly disturbed by human activity (Hviding 2003a). But were those forests pristine? Or, as asked in general terms by a team of tropical forest ecologists: how 'virgin' is virgin rainforest? (Willis et al. 2004).

My own continuing work since 1986 in the Marovo Lagoon and surrounding islands has been inextricably tied to the increasingly complicated contestations surrounding the forests of the area. When I first arrived there, I was thrown straight into the aftermath of the conflict that resulted in the final departure from the Solomons of colonial survivor Levers Pacific Timbers. That company had been operating 'scientific forestry' (which frequently seemed to be a poor disguise for agendas of clear-felling) for several decades in the Western Solomons, mainly on land obtained by the colonial administration as 'forest estate' through purchase from customary, kin-based landholding groups referred to throughout New Georgia as *butubutu* (see Hviding 2003b). But in the 1980s, only a few years after Solomon Islands gained independence in 1978, Levers Pacific Timbers operations ground to a halt after the company shifted its operations to customary land, having exhausted the tracts of 'forest estate'. Forceful local opposition was met with in northern New Georgia. As I got to know the notoriously stubborn and independently minded people of a prophet-led rural movement called the Christian Fellowship Church (CFC) who had the distinction of having made life for Levers Pacific Timbers unbearable, I quickly learned that they had not burned bulldozers, menaced company labourers (non-violently), taken a hand in provincial politics and ultimately caused the fall of this transnational giant's operations in New Georgia out of conservationist sentiments. Instead, they had done so in defence of their own autonomy as rural landholding groups, and in defiance of a colonial regime and a young national government that were seen to be conspiring with Levers to gain access to timber on customary lands in clever legalist ways aimed at bypassing the legal privilege of customary landowners (Hviding 1996a: 316–20).

Surely, if the eyes of the powerful international lobby of rainforest conservation and indigenous rights activism had been focused on the Solomons in the mid-1980s, the people of northern New Georgia, and especially the colourful and exotic CFC movement — with its powerful prophet the Holy Mama, who so well mixed traditionalism, modernist development aspirations, United States Army inspirations from World War II and singular worship practices — might have entered centre stage on the global scene as prominent tribal conservationists, perhaps alongside Amazon Indians. But the major international organisations had not then arrived in the Solomons. And the fact that New Georgians never really saw themselves as occupants of the tribal conservationist slot was amply demonstrated when in 1989 the leaders of the CFC, in their capacity as managers of vast forests, invited an Indonesian company to enter their lands and log that which Levers had left.

Since then, the northern and western parts of the island of New Georgia have seen massive land transformation through selective logging and, in designated areas subject to clear-felling, the development of large, locally owned tree plantations where *Eucalyptus*, *Gmelina*, teak and mahogany are grown in conjunction with garden crops. The development of teak plantations has been particularly extensive. But at the opposite end of the Marovo Lagoon, the large island of Vangunu has experienced a chaotic multitude of logging by many operators, and a large tract of government land on the island's south-eastern side, alienated in colonial times, has been clear-felled for the ostensible purpose of establishing a Malaysian-owned oil palm plantation which never became operative. The forests of New Georgia that in the 1980s and early 1990s appeared to outside observers as pristine, with dense canopies covering the land from coast to coast across mountain ridges, have been forever changed (Hviding 1996a; Hviding and Bayliss-Smith 2000). This process is, however, not difficult to grasp through the eyes of the Marovo people, in whose history there is a long and continuous record of transforming that which is 'wild' (*piru*) and outside the human domain into the condition of 'domesticated' (*manavasa*). Neither of these two states are regarded in Marovo as definite — thus 'wild' forest is not a timeless wilderness, but usually forest that has reverted from being 'domesticated' (Hviding 1996b).

At the edge of the 'domesticated', Marovo people have always seen new 'wild' spaces to be drawn into the former sphere by human activity. The scale of such transformation has escalated in recent and present times, though. But, as will be shown below, larger-scale permanent land transformation has increasingly become part of the local land use repertoires in the twenty-first century, as the people of Marovo Lagoon and wider New Georgia have devised new approaches to managing lands whose forest has been radically changed through logging. In the present, external and indigenous agendas for land transformation have merged, but this has taken place against a background of quite extraordinary forest modification in the past.

Forests and People in Old New Georgia: Cultural History and the Evidence of Botany

Over the past 30 to 40 years a new paradigm for rainforest ecology has been developed, in which forest disturbance is given a central role in the maintenance of botanical diversity. Gaps that are created in the forest canopy are seen as generating a range of different microhabitats, while forest tree species can be classified along a continuum from light-demanding to shade-tolerant, according to how much sun they require for regeneration. Two broad classes of tree are significant in this 'gap-phase' paradigm: pioneer species and climax species. Pioneer species germinate and establish growth only in full light in a gap after its creation, and their seedlings are not found growing below a full and undisturbed canopy (Whitmore 1984, 1990, 1998). In this view, rainforests reach their highest levels of species richness when they are in an intermediate stage of recovery from disturbance, or when disturbance is at an intermediate intensity or frequency. In these intermediate states the forest will contain the fullest repertoire ranging from pioneer species to climax species, and its biodiversity is at a maximum (Connell 1978).

How can the gap-phase paradigm be made relevant for interpreting the dynamics of forests so easily construed as 'natural' or 'virgin' in the snapshot insights of outsiders? The Marovo area lends itself well to such discussion owing to inherent contradictions. Until the early twentieth century the Marovo Lagoon, like its political counterpart and oftentimes enemy Roviana Lagoon further west, was the scene of large-scale maritime exchange and predatory warfare, as well as the hub of a regional economy based on the large-scale irrigation of *Colocasia* taro (*talo*) in pondfields (*ruta*) constructed in the forest (Bayliss-Smith and Hviding 2012, 2014a).[2] The contradictions in question are grounded in both the pre-colonial and the post-colonial. On one hand, the Marovo area has a complex history of centuries of agricultural intensification, inter-island contacts and sociopolitical transformations, yet on the other, the lagoon's catchments and its forested islands have in recent decades figured in international media as a prominent example of pristine tropical wilderness. It is well known that Marovo has on several occasions been proposed for permanent conservation, even as a possible UNESCO World Heritage site (Hviding 2003a).

The distinguished rainforest botanist T.C. Whitmore, who not only developed the gap-phase paradigm but also was the undisputed authority on the vegetation of the Solomon Islands, noted that the islands' forests below 300 metres are in many ways similar to forests throughout South-East Asia and New Guinea. However, Solomon Islands forests are lower in height, have fewer emergent trees but more epiphytes and climbers, and are impoverished in species diversity. In the forests of the Western Solomons specifically, the canopy, which is 30–45 metres tall but lower on steep slopes and at higher altitudes, is locally broken by gaps filled with climbers and thickets of smaller trees. The key to understanding some important long-term processes in these forests is a 30–50 metre tall tree called *Campnosperma brevipetiolata*. From his extensive surveys in the 1960s of forests in the British Solomon Islands Protectorate, Whitmore noted that, especially in the New Georgia Islands and on Santa Isabel, large areas of lowland forest were almost totally dominant at canopy level by this tree. Since *Campnosperma* is a light-demanding species that requires gaps in order to grow, any forest dominated by it must be the site of considerable canopy disturbance. At the time, Whitmore attributed this rather strange vegetation pattern to significant human disturbance in the areas concerned up to the end of the twentieth century, noting for vast tracts of southern Isabel where 45 per cent of large trees were *Campnosperma* that 'this particular area may well date from the abandonment of cultivations following the decimation of the local people by raiding parties from New Georgia across the Sound, which is known to have occurred within living memory late last century' (Whitmore 1969: 265). Later, his influential voice in the botanical understanding of Solomon Islands forests changed into proposing a relationship between recurrent tropical cyclones and a predominance of *Campnosperma* at canopy level.

---

2   In this paper vernacular terms are given in the Marovo language, which is a secondary doctrinal language of the CFC and of the United Church, and a primary one for the SDA Church. The latter is the numerically dominant denomination in the Marovo Lagoon. Marovo is also the field language I use. The coexistence in New Georgia of the two closely related languages Marovo and Roviana thus is strongly connected to mission history as well as to geographical distinctness, but on an everyday basis most adults in New Georgia, wherever they live and whatever church they belong to, are able to speak both languages.

This is where my own work on the political ecology of forests in the Marovo Lagoon intersected with Whitmore's observations, mediated through a mutual friend and colleague, geographer Tim Bayliss-Smith. Conversations in Cambridge led to collaborative work of longer scope. In 2000, Bayliss-Smith and I published the book *Islands of Rainforest*, where we took up Whitmore's old argument that *Campnosperma*-dominant lowland forest indicated past human disturbance. We used this insight to interpret exquisitely detailed forest-type maps produced in late colonial times, and we concluded that a canopy dominance of *Campnosperma* characterised areas that were known from oral history and from personal observations to be sites of substantial past settlement and extensive irrigated taro cultivation. Thus there seemed to be no doubt that this type of species-poor forest, so characteristic of much of New Georgia, was a messenger of widespread forest transformation up to the late pre-colonial and early colonial years. From about the end of the nineteenth century, as the systems of exchange and tribute that governed relationships between the coastal and inland dwellers of New Georgia collapsed with the end of inter-island warfare and head-hunting and with the arrival of British naval power, Christian missions and other sweeping changes, inland settlements and their associated irrigated taro pondfield systems (*ruta*) were gradually abandoned. By the late 1930s few *ruta* were left in operation anywhere in New Georgia.[3] When the coastal villagers of Marovo retreated into the bush in 1942 in fear of the oncoming war machines of Japan and the United States, pondfields were already so overgrown as to discourage any attempt at returning to growing wetland taro, and wild yams instead became the staple diet. With the abandonment of pondfields the heavily modified forests with broken canopies regenerated, but into a particular *Campnosperma*-dominated ecological configuration.

Bayliss-Smith, Whitmore and I soon embarked on a collaborative study for which we combined a range of botanical, ecological, archaeological, demographic and anthropological materials to conclude once and for all that the lowland forests of New Georgia as they appeared just before logging accelerated in the late twentieth century were not in any way 'virgin'. Instead, they were a low-diversity form of secondary growth, rich in large timber species, produced by continuous human intervention over many generations up to the late nineteenth century, or in some forests even more recently. This analysis allowed us to use the finely detailed colonial forestry maps to estimate the extent of land under dryland and irrigated cultivation around 1870–80, as well as to propose a number of demographic scenarios for the undoubtedly substantial inland populations

---

3   Wetland taro cultivation in small pondfields along tributaries of major rivers has continued on a very small scale, but as an unbroken practice, in a few locations in northern New Georgia inhabited by traditionalist people of the CFC. They are also believed to carefully preserve small intact stocks of *talo ruta* or 'pondfield taro' growing in water even where there are no pondfields. Here and there along riverbanks in these parts of New Georgia, this form of wetland taro can also be found in feral condition.

of bygone days. The population estimates made for the main area of Marovo Lagoon in the period 1870–80 gave ranges from about 12,000 to about 22,000 people, compared to the 1986 population of less than 7,000, which has increased to a 2010 population of about 15,000 (Bayliss-Smith et al. 2003: 348–51; see also Hviding and Bayliss-Smith 2000: 138–40).

Tales of the Past, Misconceptions of the Present

That the empty inner lands of the New Georgia Islands were well settled in the past is not contested. Most customary landholding groups (*butubutu*) today maintain repertoires of stories about inland settlements, ancestral shrines, ceremonial grounds and old groves of *Canarium* nut trees, framed in a genealogical perspective. Indeed, such stories are becoming all the more significant in the present-day context of inter- and intra-group disputes over customary rights to land — there is a direct correlation between recognised entitlements to customary land and the current flow of timber royalties from logging companies. In that sense, then, tales of the past inland settlement of one's landholding *butubutu* may feed straight into the realm of bank accounts for royalties and into conspicuous consumption.

Not many generations span the time from when the inland forest became disused except for occasional forays for feral pigs; most of the interior of the New Georgia Islands was not entirely abandoned until the early twentieth century. From his work in the Marovo Lagoon as a surveyor for the Royal Navy in the 1890s, Lieutenant H.B.T. Somerville reported on a cautious expedition in 1893 that ventured quite far inland and uphill on the northern part of Vangunu Island. Most surprising to the Royal Navy representative was the inland population his party encountered:

> This small expedition into the bush … points out that (1) Most of the people of this district live away from the sea coast. (2) That, however, they are probably in the habit of visiting the coast for trade etc. (Somerville 1893: 5)

In a somewhat later, more lurid missionary's tale, an account is given of how three young boys of a group of 'saltwater people', recently converted to the Seventh-day Adventist (SDA) faith (but later to become leading missionaries in their own right and tell their boyhood tale to the writer in question), experienced the inner lands of south New Georgia during a forced bushwalk in 1915 back from the mission station at Viru Harbour:

> Trekking through the dense jungle, the three youths crossed parts of New Georgia, carefully bypassing strange villages, devil-dancing grounds, and piles of sacred spirit stones in the forest groves. They were out of their territory and among

enemies. They had no desire to provoke the wrath of the people or their spirit-gods. Happily they found the coast and from a friendly old woman borrowed a canoe and paddled across the lagoon to their own shores. (Were 1970: 26)

As boys of a coastal group with little experience of life in the interior, they are likely to have been acutely aware of the strange world they encountered. The continuing existence of a number of interior settlements in Marovo and in northern New Georgia as late as 1910–15 is in line with the conventions of oral history in the area, as well as with some eyewitness accounts by missionaries and travellers.

But in most of New Georgia the interior had been depopulated by the turn of the century. As the inner lands were abandoned by all those people indicated by our population estimates — numbers that may have been decimated because of epidemic diseases known to have had locally devastating consequences from around 1870 — old settlements and finely engineered pondfield systems for taro were left to the encroaching forest. Concurrently a new form of land transformation took place from the early years of the twentieth century in areas closer to the lagoon coast. This was the very first intervention of the global political economy into the forests of the Western Solomons, and it preceded the logging boom of the 1990s by almost 100 years (Hviding and Bayliss-Smith 2000: 150–9). The rapid build-up of a coconut economy throughout New Georgia caused large areas of shoreline forest, sago palm stands and mangrove vegetation to be cleared and replaced by coconut groves. The hillsides closest to the growing coastal villages were now placed under intensive shifting cultivation of food crops, and the only interior forest features still receiving attention from a people now fully coastal were old groves of *Canarium* nuts, still tended through the regular removal of undergrowth and invading climbers. The massive tracts of remaining inland forest were simply left to regenerate, with the inevitable dominance of *Campnosperma brevipetiolata* in many of those areas that had been most heavily modified by centuries of pondfield construction and maintenance. This, then, was the forest scene encountered by loggers and conservationists alike in the late twentieth century — a robust-looking forest with a dense canopy and an abundance of tall trees festooned with climbers and orchids, seen by the former as timber-rich and by the latter as pristine/primary/virgin. Gregarious creepers, climbers and herbaceous forest floor shrubs hid most signs of past human habitation from those not in the know, but for locals who did know, the history of human habitation and forest modification were there to see.

Quite a different, and markedly less dynamic, interpretation has been presented by agents of forest conservation, who have seemed to be reluctant to consider anthropogenic disturbance as a ubiquitous feature of the forests in places like

Solomon Islands. For example, as the logging boom was gaining momentum, a New Zealand conservation organisation presented its case to the Australian government, the principal aid donor to Solomon Islands, as follows:

> The proposed protected area [i.e., the forests of Marovo Lagoon] has been little disturbed: hurricanes are infrequent and of low intensity, and human impact has been limited to gardened land and some timber removal on Vangunu and Nggatokae. The result is a region of extensive primary forest cover, including undisturbed forest community transitions from the sea to mountain tops. (Lees et al. 1991: 95)

Let this quote stand as representative for an environmentalist view not informed by either cultural history or by rainforest ecology. A contrasting view of those same forests not as 'undisturbed' at any time would focus instead on their ecological resilience under successive stages of transformation, not only after centuries of regular disturbance for settlement and agriculture, but even, and notably so, in the face of modern logging operations. Observations by Tim Whitmore in the mid-1990s in fact indicated successful regeneration of logged forests, even in northern New Georgia where clear-felling by Levers Pacific Timber had resulted in a long delay of regeneration as degraded areas were invaded by creepers such as *Merremia*. Fourteen years after this most severe disturbance, *Campnosperma* and *Pometia* trees had pushed their way through the carpet of creepers to reach a height of 13 metres (Whitmore 1995). Whitmore argued that a succession back to mixed forest seemed to be assured given no additional human disturbance. It is a fact that, unlike Levers, few of the Asian companies operating in New Georgia since the 1990s have had agendas of clear-felling. Instead they have preferred to work over often volatile logging concessions on customary land in a quick and often confused manner, crisscrossing the land with dirt roads and skid tracks and ripping out as many valuable trees as possible before any trouble with landowners arises. Ironically, this rough and rapid approach leaves far more vegetation standing than the scientifically informed clear-felling carried out by Levers. Total forest loss through the planting of oil palms in logged areas remains a sinister possibility, though, and was indeed envisaged for the notorious tract of government land on Vangunu clear-felled in the 1990s by Silvania Products of Malaysia, although it never happened. More diverse post-logging trajectories have been evolving in local hands.

Transforming the Forest: Logging, 'New Life' and Multiple Trajectories

As the twentieth century drew to a close, new scenarios were to be developed in the historical and political ecology of New Georgia's forests. While logging by a bewildering range of companies in central Marovo proceeded in a chaotic fashion, creating scars in both landscape and social life, the northern areas under control by the CFC, once the bane of Levers Pacific Timbers, followed a different course. From 1989 until today, logging in the area has with few exceptions been carried out by only one company — Golden Springs International of Indonesia (where it operates as Sumber Mas, of which 'Golden Springs' is a translation). Golden Springs operations have been managed in close consultation with the leaders of the CFC, whose organisation functions as a corporate tool for combining the control over several dozen tracts of customary land throughout northern and western New Georgia held by the constituent kin groups of the movement. The encompassment of church leadership, since the days of the Holy Mama, of land rights vested with a large number of individual *butubutu* throughout New Georgia, constitutes the movement's main corporate power base (Hviding 2011: 9–10). Although exact details of the agreements between company and church are elusive given the secretive nature of the CFC, it is a well-known fact that its royalty rates are far higher than for any other logging operation in the Western Solomons, and probably the highest anywhere in Solomon Islands. It is my impression that the willingness of Golden Springs International to provide substantially higher royalty rates over several decades is directly tied to the benefits the company receives from its extraordinarily long-term presence in the land.

Leaders and adherents of the CFC, as well as informed non-CFC observers, make no bones about the considerable wealth accumulated by this movement over the years. Indeed, wealth accumulation is an integrated aspect of a doctrinally central concept of 'New Life' to be attained for followers. The financial system of the CFC manages flows of logging-generated revenue from the periphery of customary landowners (on whose land logging operations take place) to the centre of church leadership. This accumulation is a centralised one that keeps timber royalties virtually outside the village economy and instead supplies villages with services such as schools, health facilities and transport. This can be seen as a new twist on old Oceanic patterns of chiefly redistribution, and was a strategy not least significant at the dawn of the twenty-first century when the already meagre government services to rural areas were disrupted during the so-called 'Tension' — a condition of limited civil war (mainly on the island of Guadalcanal) leading to a collapse of both government and national economy (Moore 2004; Allen 2013). More specifically, the CFC consolidates fragmented kin-based authority over land by collectively organising customary landowning groups in large-scale ways that lack parallels elsewhere in the Solomons. A range of self-initiated economic development projects are operated by the various arms of the CFC organisation, through an approach that avoids collaboration with non-governmental organisations and refuses most outside aid, instead relying on its own funding in the form of accumulated logging royalties.

With its somewhat peculiar forms of worship, and its insistence that its leader is the earthly manifestation of the Holy Spirit, the CFC has, since it was established through a breakaway from the Methodist Mission in the 1950s, been regarded with distrust by many Solomon Islanders. Thus knowledge of its singularly successful way of adapting to the logging boom remained limited. The public image of the CFC in Solomon Islands was improved and national-level knowledge of the movement expanded with the unexpected media coverage given in 2005 of two high-level visits to the CFC headquarters at Duvaha in North New Georgia, first by the prime minister, then by the governor-general. Further national media attention was given later that year with the bestowment of a knighthood by Queen Elizabeth II upon the CFC's spiritual leader, the Reverend Ikan Rove, who visited London for the occasion. Known to CFC adherents as the Spiritual Authority, Reverend Rove is the son of the Holy Mama, who founded the movement and brought it into a recognised state as an independent, indigenous church by 1960. Since succeeding his father, Reverend Rove has seen it as his foremost mission to promote economic development for his people, moving them into a new era he has designated the 'time of development' (*kolokolo divelopmen*). In line with this, the large-scale development by the CFC since 2000 of huge tree plantations on customary land logged and clear-felled by the Golden Springs company, particularly the communal planting of what must be hundreds of thousands of teak trees, has caught attention in the national search for post-conflict options for rural development, although being seen as somewhat of an enigma by foreign proponents of 'community development projects' (e.g., Makim 2002; see also Hviding 2011).

In the Western Solomons, environmental histories of forests are, as will be clear by now, closely connected to the different paths followed by the three church denominations of these islands: the CFC, the United Church (formerly the Methodist Mission), and the SDA Church.[4] When the Asian logging boom hit Solomon Islands in the early 1990s, the CFC was ready to take the fullest advantage of its already existing corporate system of amalgamated customary

---

4   See Hviding (1996a: 118–24) for mission history and more recent developments in Marovo Lagoon and McDougall (2008) for more general comparison of the potential roles of different denominations for generating rural development in the Western Solomons. Hviding (2011) gives a comprehensive analysis of the history of the CFC and the movement's most recent accomplishments. The account that follows is based on seven separate field visits throughout New Georgia between 2003 and 2012.

land under centralised authority. As logging accelerated, many SDA groups of central Marovo and northern New Georgia found themselves at an advantage as major landowners through their long-term history as 'bush people' whose previous generations had lived and worked in the interior. In contrast, the original strongholds of the SDA faith in the Solomons remained largely landless, determined by their history as sea- and reef-holding coastal or 'saltwater' groups of central Marovo Lagoon, easily contacted by the pioneer SDA missionaries. As logging gained momentum, the inequalities among New Georgia's SDAs in terms of landholdings, coupled with the bilaterality of kinship throughout Marovo and Roviana, generated massive land disputes that were further deepened by the fact that 'saltwater' people with limited primary rights to land remained the most influential in church matters. Nevertheless, a great number of logging operations commenced on SDA land, but with the majority of landholding *butubutu* having no' resident chief (since most younger chiefs had become urban elite), a lack of planning has resulted in a chaotic picture where almost every SDA *butubutu* with significant landholdings has engaged its 'own' logging company. This chaos is inscribed on the SDA lands of central Marovo — known as the Bareke hills of northern Vangunu — where a dense and confused pattern of logging roads and shipment ponds mar the hills and coastlines. In 2007 I observed approximately one log shipment pond for every 4 kilometres of coastline in central Marovo, whereas in the CFC areas of northern Marovo the log shipment ponds were up to 20 kilometres apart and the logging road network was much more orderly.

A similar contrasting picture characterised the financial side of logging. I have already described the CFC system of keeping timber royalties out of the village economy, centralising and accumulating the funds, and carrying out regular redistribution to church members in the form of subsidies of school fees, tax obligations and rural infrastructure. Among the SDAs, timber royalty rates have generally been lower than those of the CFC, owing to the poor bargaining situation of fragmented *butubutu* and weak internal leadership. Still, a large number of logging operations on SDA land have generated substantial income, much of which has been disputed through court cases. Whatever the outcome of these cases, the final recipients of royalties among the SDA groups have spent most of the money on consumption goods, short-lived investments such as outboard motors, and a profusion of not always durable houses of permanent materials. Apart from a number of very impressive church buildings now standing more or less finished in villages of central Marovo (reflecting an emphasis among Adventists and Methodists on large churches as rural status symbols), few enduring results are to be seen today from 10–15 years of huge windfall profits from logging. The lands are devastated by the unplanned use of logging machinery, and subsistence gardening suffers. Typically, the SDAs (who for sectarian reasons have always refused to have government-operated schools) still have to pay substantial annual school fees for their children at primary and secondary schools, and the quickly evaporated royalties have not assisted families in this arduous task.

While the Methodist (or United Church) villages of Marovo and of New Georgia more widely all have government schools and so are not burdened with school fees at the level of the SDAs, they have otherwise lost out on most scenes of contemporary political economy in the Western Solomons, dependent as these scenes are on the attainment and management of windfall profits from logging. Geographically squeezed in between the huge landholdings of the CFC to the north and of the 'bush' SDAs in central Marovo, United Church groups with their more limited lands were disadvantaged in concrete terms, while most of their villages shared the problems of erosion of customary authority shown among the SDAs. In the Roviana Lagoon area of western New Georgia problems were centred on the polarisation between an economically declining United Church and a financially expanding CFC, once a detested breakaway movement. The level of land disputes among United Church people has been staggering, nowhere as much as in the old heartland of the church in west-central Roviana. As among the SDAs, timber royalties have been largely wasted on consumption and short-term benefits. In the most recent years, the United Church has been hit even harder by the emergence of a new breakaway movement that claims to represent the true Wesleyan faith.

And so the forests and the land, the people, their social life and rural economies have been transformed in recent decades all over the Western Solomons, as in so many other parts of Melanesia. What is notable from the vantage point of New Georgia is the way in which three very different church dominations, each of which pervades most dimensions of everyday life in its adherent villages, have followed strongly diverging paths in relation to the complex local–global entanglements of the forests that grow on their customary land. The diversity of strategies and outcomes from both local and global agents defy any simplistic image of rural Solomon Islanders as passive victims of ruthless foreign companies, although invariably the Asian companies make by far the most money from the trees that are felled and exported as round logs.

The New Anthropogenic Landscape and the 'Time of Development'

The ragged and chaotic post-logging condition of much of central Marovo has certainly prevented the place from ever again approaching World Heritage listing, and the efforts of international conservationists in the area seem to have been without much result. There is much current uncertainty among the SDA and United Church villagers of the area as to the future of the forest. Perhaps the likely scenario is one of gradual regeneration through disuse and neglect — a somewhat extreme case of Whitmore's gap-phase paradigm, made more difficult by the environmental destruction that has taken place through the compacting and erosion of soil. This is not so in the northern areas under CFC control, where the government's Standard Logging Agreement (in most cases blatantly ignored by companies once signed) has been largely adhered to owing to the close connections between companies and church leadership: forest near water catchments is left undisturbed, no logging is carried out on high slopes, and roads are gravelled.

More significantly, the forest landscape that is now emerging after logging in the areas controlled by the CFC represents a new stage in New Georgia's anthropogenic forest; another transformation across the *piru–manavasa* ('wild–domesticated') continuum. On both sides of logging roads that run parallel to the coast a few hundred metres inland are family-operated blocks of tree plantations where some food crops are also grown in interspersed fashion. These are the lands now known throughout the Solomons to support huge, rapidly growing stands of teak, the valuable harvest of the future, and from the point of view of the leaders and working people of the CFC the fast-growing tree plantations amount to a sustainable enrichment of customary land. That argument does not go down well with conservationists who point out that lowland rainforest was actually removed to make room for the stands of teak, but those arguments do not go down well with the people of the CFC. It has in fact been pointed out to me in conversations with eager tree planters of the CFC that this as good as permanent transformation of parts of the lowland forest into 'something more useful' is not unlike the permanent 'domestication' of inland forest that took place in the old days, when gentle slopes near streams were cleared and became the sites for large assemblages of stone terraces for taro irrigation.

Two key locations of the CFC today display the highest degree of planned land and forest transformation anywhere in the Western Solomons. In a sheltered inlet on the exposed Kusaghe coast of north-west New Georgia lies Paradise, a large village founded in the 1950s by the Holy Mama, on the grounds of a former village called Menakasapa. Paradise was for several decades the very demonstration site for the CFC's successful and disciplined communal life. Houses there are still organised in orderly rows, and somewhat stricter regulations apply to everyday life here than elsewhere within the CFC movement. The flat coastal land around the village and inlet is fully planted with coconuts. When travelling the choppy seas of the 'Kula Gulf' between Kusaghe and Kolobangara, a sweeping view can be had of the slopes behind Paradise, which are totally planted with well-circumscribed sectors of *Gmelina*, *Eucalyptus* and teak trees (Figure 3.2). This is where, over a few days in 2000 in a now legendary event, hundreds of people cleared the entire hillside of residual post-logging vegetation and rapid regrowth, then planting thousands of tree saplings. In this capacity Paradise entered the twenty-first century as the CFC's demonstration site for the new era of tree plantations, although no longer the headquarters of the movement.

Figure 3.2 Tree plantations at Paradise village: *Gmelina*, *Eucalyptus* and teak in distinct blocks.

Source: Photo by Edvard Hviding, 2005.

The Paradise event, with minor modifications, has informed commonplace procedure for the CFC when transforming hillside rainforest to tree plantations. Usually, to take advantage of machinery, the logging company operating in an area destined for a new plantation is requested by the CFC to fell more than its usual repertoire of trees, and to clear away debris with bulldozers. The large-scale communal clearing and planting, with the participation of church members from a wide area, then becomes a rapid process, after which the local landholders will have the task of maintaining the growing trees and keeping the plantations clean from encroaching vegetation. Often, to maximise usefulness, new hillside tree plantations behind CFC villages combine as food gardens with the dense growth of selected garden crops (sweet potato, cassava, pineapple, etc.) protecting the plantation soil from the blazing sun.

From about 1990 the CFC has had its headquarters at Duvaha, the site of a large primary boarding school, on the mangrove-fringed lagoon coast of North New Georgia, and centrally placed in the Kalikolo area, which is the CFC's historical heartland (Figure 3.3). In and near Duvaha are several large modern houses belonging to the CFC leadership, and behind the village area at an elevation of about 50 metres are the former logging roads lined with lush tree plantations about 100 metres to each side, divided into family-held *bloko* or 'blocks'. The children who attend the boarding school come from several villages and hamlets throughout the Kalikolo and Kusaghe areas, and their families take turns to live at Duvaha and care for the entire school through food production, cooking and communal work. While there, they also tend to allocated work in the tree plantations, whether their own or those of relatives. Today's CFC strongholds like Paradise and, particularly, Duvaha, demonstrate a level of work organisation and a level of development ambition and leadership allegiance echoing Methodist Mission strongman Reverend J.F. Goldie's 'industrial mission', in which New Georgians were taught 'that the development of their land is a Christian duty and that labour is honourable' (in Bennett 2000: 45). Given that the Holy Mama claimed to have inherited the Holy Spirit from Goldie, these connections in thought and practice are not surprising — but today's 'development' is firmly in local hands.

Figure 3.3 'The Time of Development': Duvaha, North New Georgia.
Source: Photo by Edvard Hviding, 2012.

Duvaha also demonstrates other approaches to recreating the land. A few hundred metres along the mangrove-fringed coast is one of the large log shipment ponds of the Golden Springs company, in all its grimness visualising the intimate relationship between church and company in the Spiritual Authority's designated 'time of development'. Further ahead, accessible through a crocodile-infested mangrove channel, footpaths lead to the most recent land transformation by the CFC at a place called Kelehae: a huge food garden for the Duvaha population, covering all the slopes of a low, circular crater-like basin that is hidden from the lagoon by a low coastal ridge. This singular topographical form, until recently covered by coastal forest interspersed with some small swiddens, has been fully cleared and thoroughly domesticated under the leadership of the former provincial secretary of the Western Solomons, Mr Narcilly Pule. This massive effort has produced a food garden of a scale not seen in New Georgia for a very long time, echoing the so-called *chigo lavata* or 'huge communal swiddens' cultivated by the bush dwellers of the old days. Integrated with irrigated *ruta* pondfields in a continuous 'wet-and-dry' cultivation system, *chigo lavata* were used to build up large surpluses of yams and dryland taro for cyclical feasts. At Kelehae, the radically transformed land reflects the CFC's revival of selected practices of the past.

The impressive rural machinery of the CFC has been significantly altered in recent years through the long illness of the Spiritual Authority. Having suffered from several strokes, he remained largely immobile and unable to speak after 2010, leading a secluded yet itinerant life on board his ship MV *Silent One*, a former research and live-aboard diving vessel of 30 metres which he purchased in 2008 (reportedly for SBD 3 million). When he died in June 2014, several years had passed during which his sons and circle of advisers had been locked in conflict over the future religious orientation, leadership and financial management of the church with the Spiritual Authority's surviving siblings, led by his younger brother the Hon. Job Dudley Tausinga. In the process, the close-knit CFC villages throughout New Georgia split into two factions, as the central authorities of the church expelled Tausinga, his sons and the other siblings of the disabled Spiritual Authority, justifying this radical move with irreconcilable doctrinal differences as Tausinga promoted a message of reformed leadership, with less of the elevated status attained by the Spiritual Authority (who was referred to increasingly in the Roviana-speaking areas as *Tamasa Toana* or 'Living God'). While the split and its consequences for the cohesion of the CFC, and the complicated matter of succession, are being sorted out, the CFC's many projects have proceeded at a somewhat slower pace, although the tree plantation *bloko* allocated at family level are still tended in the anticipation of the future value of teak, although who tends what may no longer be a clear-cut case in villages where the split has led to internal division, and some plantations are likely falling into disrepair.

The Hon. J.D. Tausinga, one of the key actors on the New Georgia forest scene ever since logging started there, was the member of parliament for North New Georgia continuously from 1984 to 2014; the country's longest-serving parliamentarian ever. His illustrious political career includes the cabinet portfolios of foreign affairs, natural resources, education, and forestry, as well as deputy prime minister, several terms as opposition leader, and an almost successful bid for prime minister. At the time of his defeat in the 2014 elections (to a candidate of the other faction of the CFC), he was the deputy speaker of parliament. Meanwhile, he has been operating international business enterprises, including the non-profit import and sale of Chinese-made replicas of Yamaha outboard motors branded 'Tausinga'. The university-educated Tausinga is the youngest son of the Holy Mama and retained the CFC's presence in provincial and national politics since the days when he himself had an instrumental role in the heated political processes and local-level resistance that ultimately led to the withdrawal from Solomon Islands of Levers Pacific Timber. Outside observers were surprised when in 1989, this influential, for some notorious, person thought to be an anti-logging activist brought in a new company. In a 1997 overview of the Solomon Islands logging scene, Ian Frazer (1997: 56) commented how the anti-logging movement in the country 'suffered a major setback when Tausinga unexpectedly changed his views and threw his support behind a new foreign investor in North New Georgia, the Indonesian-based company Golden Springs International'. From the local point of view, however, Tausinga's actions were not seen in any way as an unexpected change of views, but as a fulfilment of the ambitions of the CFC to choose their own collaborators and then dictate the terms of collaboration.

From Tausinga himself comes a telling summary of development strategy in the form of a discussion paper entitled 'Economic Concept'. The paper was presented to a gathering of Western Solomons political and administrative leaders in May 2002, at the height of the 'Tension' when the Solomon Islands government was bankrupt and defunct without any rural reach, and the Western Solomons was verging on secession. Tausinga's paper can be seen as a succinct expression of CFC policy on resource development; it was intended as a substantial intellectual input into the development policy of a new state of Western Solomons and it echoes the author's many-sided personal experiences and his wide readings that include Greek philosophy, Oriental religions and Marxist economics:

> The guiding principal of economic concept and plan is encapsulated on the unwritten wisdom of the spirit of the modern world and thus:
>
> Resources are not resources until you develop them. (Tausinga 2002: 2)
>
> The exercise we take in identifying the economic opportunities shall encourage both local and foreign investors to participate in commercial activities. (ibid.: 5)

The incorporation of thoughts, objects, persons and institutions from outside is a foundational strategy in New Georgian ways of relating to wider worlds. And so Tausinga's accentuation of the complementary roles of local and foreign investment defines an important space between the globalised challenges of today and the situation in the old days, when the predatory approach to inter-island relationships of New Georgia's polities exploited the foreign of the day. The notion that 'resources are not resources until you develop them' may seem drawn from Marxist thought, but accurately reflects certain New Georgian attitudes that come to the forefront when villagers and their leaders engage in discussion with conservation organisations (Hviding 1996a: 56–7, 365–6; 2003a). This prevailing attitude is not always well understood, and certainly expresses something quite different from an indigenous ethics of conservation. Indeed it has to do so, in an ontological context where there is no nature as distinct from culture, but simply a continuum between what is 'wild' (*piru*) and 'domesticated' (*manavasa*).

The observations I have made in Marovo and around New Georgia and the interpretations I have made over the years are those of a rural Melanesian people pursuing agendas of autonomy over an environment on which their everyday lives are based. That autonomy also includes, as reflected in Tausinga's paper, a local insistence that 'resources' may be identified as and when required, such as in the case of lowland forest being converted into (first) monetary value and (second) what is envisaged as perpetually yielding plantations of valuable hardwoods. In any case, those forests were never pristine. The 'wild' forest that is rapidly being 'domesticated' in New Georgia today is actually itself a product of regeneration, through which the domesticated forests of former generations reverted into a wilder condition. An argument is made for viewing Melanesian forest history in the longest term possible, in order to highlight the fact that present-day forest transformations may not be best understood through ideas of radical change, devastation and destruction, but, like in the New Georgia example, as complicated oscillations between more or less stable, locally defined states that engage with levels of larger scale.

References

Allen, M., 2013. *Greed and Grievance: Ex-militants' Perspectives on the Conflict in Solomon Islands, 1998–2003*. Honolulu: University of Hawai'i Press.

Balée, W. (ed.), 1998. *Advances in Historical Ecology*. New York: Columbia University Press.

Bayliss-Smith, T. and E. Hviding, 2012. 'Irrigated Taro, Malaria and the Expansion of Chiefdoms: *Ruta* in New Georgia, Solomon Islands.' In M. Spriggs, D. Addison and P. Matthews (eds), *Irrigated Cultivation of* Colocasia esculenta *in the Indo-Pacific: Biological, Social and Historical Perspectives*. Osaka: National Museum of Ethnology (Senri Ethnological Studies 78).

Bayliss-Smith, T. and E. Hviding, 2014a. 'Taro Terraces, Chiefdoms and Malaria: Explaining Landesque Capital Formation in Solomon Islands.' In N.T. Håkansson and M. Widgren (eds), *Landesque Capital: The Historical Ecology of Enduring Landscape Modifications*. Walnut Creek (CA): Left Coast Press (New Frontiers in Historical Ecology).

Bayliss-Smith, T. and E. Hviding, 2014b. 'Landesque Capital as an Alternative to Food Storage in Melanesia: Irrigated Taro Terraces in New Georgia, Solomon Islands.' *Environmental Archaeology* advance article. DOI: 10.1179/1749631414Y.0000000049.

Bayliss-Smith, T., E. Hviding and T.C. Whitmore, 2003. 'Rainforest Composition and Histories of Human Disturbance in Solomon Islands.' *Ambio* 32(5): 346–52.

Bennett, J.A., 2000. *Pacific Forest: A History of Resource Control and Contest in Solomon Islands, c. 1800–1997*. Cambridge: White Horse Press and Leiden: Brill Academic Publishers.

Blaikie, P. and H. Brookfield, 1987. *Land Degradation and Society*. London: Methuen.

Connell, J.H., 1978. 'Diversity in Tropical Rain Forests and Coral Reefs.' *Science* 19: 1302–10.

Frazer, I., 1997. 'The Struggle for Control of Solomon Island Forests.' *The Contemporary Pacific* 9(1): 39–72.

Hviding, E., 1996a. *Guardians of Marovo Lagoon: Practice, Place, and Politics in Maritime Melanesia*. Honolulu: University of Hawai'i Press.

Hviding, E., 1996b. 'Nature, Culture, Magic, Science: On Meta-Languages for Comparison in Cultural Ecology.' In P. Descola and G. Pálsson (eds), *Nature and Society: Anthropological Perspectives*. London: Routledge.

Hviding, E., 2003a. 'Contested Rainforests, NGOs, and Projects of Desire in Solomon Islands.' *International Social Science Journal* 178: 539–54.

Hviding, E., 2003b. 'Disentangling the *Butubutu* of New Georgia: Cognatic Kinship in Thought and Action.' In I. Hoëm and S. Roalkvam (eds), *Oceanic Socialities and Cultural Forms: Ethnographies of Experience*. Oxford: Berghahn Books.

Hviding, E., 2011. 'Replacing the State in the Western Solomon Islands: The Political Rise of the Christian Fellowship Church.' In E. Hviding and K.M. Rio (eds), *Made in Oceania: Social Movements, Cultural Heritage and the State in the Pacific*. Wantage (UK): Sean Kingston Publishing.

Hviding, E. and T. Bayliss-Smith, 2000. *Islands of Rainforest: Agroforestry, Logging and Ecotourism in Solomon Islands*. Aldershot: Ashgate.

Lees, A., M. Garnett and S. Wright, 1991. *A Representative Protected Forests System for the Solomon Islands*. Nelson (New Zealand): Maruia Society.

Makim, A., 2002. *Globalisation, Community Development, and Melanesia: The North New Georgia Sustainable Forestry and Rural Development Project*. Canberra: The Australian National University, State, Society and Governance in Melanesia Program (Discussion Paper 2002/1).

McDougall, D., 2008. *Religious Institutions as Alternative Structures in Post-Conflict Solomon Islands? Cases from Western Province*. Canberra: The Australian National University, State, Society and Governance in Melanesia Program (Discussion Paper 2008/5).

Moore, C., 2004. *Happy Isles in Crisis: The Historical Causes for a Failing State in Solomon Islands, 1998–2004*. Canberra: Asia Pacific Press.

Robbins, P., 2004. *Political Ecology: A Critical Introduction*. Oxford: Blackwell.

Somerville, H.B.T., 1893. 'Report Concerning the Bush in the Vicinity of Lihihina Island, Marovo Lagoon, New Georgia, Solomon Islands.' MS report addressed to Commander A.J. Balfour, H.M.S. 'Penguin'. London: Balfour Collection, Royal Geographical Society.

Tausinga, J.D., 2002. 'Economic Concept: A Discussion Paper at Leaders' Meeting, Gizo, Western Solomons 02–04 May 2002.' Unpublished paper.

Were, E., 1970. *No Devil Strings*. Mountain View (CA): Pacific Press Publishing Association.

Whitmore, T.C., 1969. 'The Vegetation of the Solomon Islands.' *Philosophical Transactions of the Royal Society B: Biological Sciences* 255: 259–70.

Whitmore, T.C., 1984. *Tropical Rain Forests of the Far East*. Oxford: Clarendon Press (2nd edition).

Whitmore, T.C., 1990. *An Introduction to Tropical Rain Forests*. Oxford: Clarendon Press.

Whitmore, T.C., 1995. 'Environmental Assessment of Proposed Tree Plantations on New Georgia by Solomons Sustainable Products Ltd.' Report for Commonwealth Development Corporation, London.

Whitmore, T.C., 1998. *An Introduction to Tropical Rain Forests*. Oxford: Oxford University Press (2nd edition).

Willis, K.J., L. Gillson and T.M. Brncic, 2004. 'How "Virgin" Is Virgin Rainforest?' *Science* 304: 402–3.

4. Forests of Gold: From Mining to Logging (and Back Again)

JAMON ALEX HALVAKSZ

From here, if you walk towards the Watut, there is timber.[1] If you go towards Biaru, there is timber. If you go towards Garaina, there is more timber. If you walk down the Markham Valley, towards Madang, you follow the timber. Even in the Highlands, if you look, there is some timber there … Wau is the true origin of this timber, Wau is its home. (Elauru resident, 2001)

Since the early days of the 1920s gold rush, the upper Bulolo Valley, Morobe Province, Papua New Guinea, has been recognised as a potential source of timber. Early miners used trees to create their camps, fuel their fires, and support the work of panning and mining. As corporate mining took over the principal fields, timber became its backbone. And even as they mined, the expatriate mining community was well aware that the timber had value, that if they could find a route to the coast, the giant klinkii (*Araucaria hunsteinii*) and hoop pines (*Araucaria cunninghamii*) might win their weight in gold. The mining community and government officers were so sure of the prospect that they pushed for a road and, barring a road, an aerial tramway to haul timber down to the coast (Lane Poole 1935). Throughout this expansion, Biangai speakers living along the upper end of the Bulolo Valley surrounding the colonial township of Wau have remained on the margins. Many different groups benefited from early gold and timber extraction, but few were Biangai.

---

1   In Tok Pisin, the term *plang* was used. This has a number of references, the first being a plank of sawn timber. Here I use the second, referencing timber trees.

Today, in spite of their political and economic marginality, the Biangai continue to maintain their centrality in the circulation of timber. As noted in the epigraph above, some Biangai claim that Wau is the home of these giant pines, in much the same way that its soils are the source of yams and gold (Halvaksz 2008). In making this claim, my Biangai friend is placing Biangai landscapes at a central node in the global market for timber and other resources. For him, this is a *true* network — part of a system of gift exchanges between ancestors, Biangai and global markets[2] — one with ongoing relationships, transactions and, most importantly, obligations. If nothing else, it reaffirms that it is their timber to use.

This chapter examines the efforts of Biangai villagers, Morobe Province, to participate in a local timber market, the relationships that they require and the obligations that they engender. After presenting the broader historical context of Biangai associations with gold mining and commercial forestry, I focus on recent efforts by Elauru villagers to engage in community-based timber production (cf. Salafsky et al. 1997/8).[3]

Making Forests in the Wau–Bulolo Valley

The only asset that I can see that will replace the wasting product, gold, is timber. The region is purely a forest one, and its climate is decidedly a forest climate ... In short, it is proven forest country. (Lane Poole 1935: 11)

Throughout the boom and bust cycles of gold and the rise and fall of commodity prices for coffee, timber has remained a viable resource. While Lane Poole foresaw a time in which timber would replace 'the wasting product, gold', timber and practices of logging in the area have instead offered some stability through macro and microeconomic transformations. Thus, the value of forests along the upper Bulolo are very much made through the colonial and post-colonial practices of mining, as well as local work of supporting and resisting those efforts. The forests, as such, do not stand apart from the gold.

---

2   Jacka (2001) describes a similar set of obligations deriving from the land itself. Seen as an inalienable possession, the land upon which the mining town of Paiam is situated establishes a set of obligations between the company and Ipili community. Likewise, Biangai conceptualise the land as generative of a seemingly renewable resource, trees for timber. However, in contrast to land, timber is alienable. Still, the exchange of timber retains obligations for reciprocal relationships.
3   Salafsky et al. (1997/8) distinguish community-based timber production from other more inclusive community or social forestry efforts to highlight project emphasis on local timber over non-timber forest products, fuel wood and food production. Community-based timber production is also distinct from industrial timber harvesting as the former emphasises local residence and ownership, and efforts to add value at the point of harvest are less capital intensive and involve local reinvestment.

Analytically, this raises important challenges to existing literature on the construction of state power around forest resources (e.g., Sivaramakrishnan 1995; Li 2007; Winkel 2010). As Sivaramakrishnan (1995) argued, colonial forestry can be as much about knowing the forests as creating systems of knowledge about the peoples that inhabit them. In India, he argues that such systems of meaning continue to inform contemporary practices, creating government technologies for controlling people and resources. Thus,

> While forest dwellers were being sorted into types by tribe and caste, the forests themselves were arranged in categories by dominant genera and species. Such description and the laying down of taxonomic structures to represent biotic communities presaged colonial development projects in which human and natural resources were harnessed for imperial purposes. (Sivaramakrishnan 1995: 6)

The colonial project, therefore, organised and controlled people and trees, but also generated discourses of management and conservation that continue into the present. Winkel (2010: 85) notes this point in reviewing the use of Foucauldian concepts in forest policy: 'many colonial forest policy analyses carried out in developing countries also emphasize the continuing prevalence of colonial practices in the postcolonial areas'.

Likewise, Li (2007) has detailed how forest policy shaped colonial and post-colonial relations in Indonesia. However, she extends her discussions to consider the agentive possibilities of community forestry in the assemblage of diverse interests (official forestry departments, conservationists, agroforestry experts, legal advocates, activists, donor agencies and forest villagers). This assemblage is conceptualised as a generative space where diverse 'lines of flight' (Deleuze and Guattari 1987) make different articulations of human agents possible. Forestry practices engage complex networks in and of themselves. But placing them into a wider mineral extraction and agricultural context adds layers to these networks, and reveals discontinuities in state power, as well as possibilities for local empowerment.

During the 1920s and 1930s, when gold was driving the development of Wau and Bulolo (Halvaksz 2006, 2008; see also Sinclair 1998; Waterhouse 2010), forestry mobilised different agencies, even as forest policy was used as a technique for control over both expatriate and indigenous communities. Importantly, government foresters aspired to regulate the exploitation of forests by miners. Aside from revealing discontinuities in the implementation of colonial state power, the history recounted below informs ongoing practices of community forestry today.

Like other areas under colonial control, local populations were accused of deforesting the region through their 'wandering habits and hunting fires' (Taylour and Morley 1933). In a presentation to the Australasian Institute of

Mining and Metallurgy, Taylour and Morley (as mining warden and assistant mining warden for the Morobe goldfields respectively) noted that the district was 'well supplied with a variety of good timbers' (ibid.: 22), with sawmills 'scattered throughout the main centres of the goldfields' (ibid.: 23). They described miners as making use of the readily available timber resources for housing and mining.

In reading early accounts of the goldfields, the availability of timber is often mentioned either directly as a resource for the industry (as above), or indirectly through descriptions of life on the goldfields. For example, in her account of life in the early days of the goldfields, Doris Booth describes the use of timber by prospectors:

> The boys cut down a tree, from which they took a length of about six feet. Then, with an Adze, they chipped out the middle of it, leaving the bottom and sides, and making something like a native canoe. This was about one foot wide and six feet long. When this was completed, they made riffles out of lawyer cane — an oblong frame of network of sticks, like a sieve. This was placed on the bottom of the box, the layers of cane being as close as possible. Then a flume like a wooden pipe was made for the water, which was diverted from the creek into the box. (Booth 1929: 87)

Early on, part of the capital for the emerging mineral industry was created from the forests. Accounts of the goldfields like Booth's express little concern about local timber rights, although government regulations did exist. Thus, timber, like land and water, was intimately linked to the production of gold.

The potential for a productive timber industry was recognised early on. However, it was forestalled, at least in part, because of local demand by the mining industry, but also because of transport concerns. The logistics and costs of building a road to what was then the main north coast port of Salamaua were especially daunting. A slow colonial bureaucracy, perhaps more familiar with mineral claims than timber concessions, also delayed the development. As Charles Booth lamented in a 1939 letter to *Pacific Islands Monthly*, he had tried for 13 years to gain a licence for 25,000 acres, noting 'When I arrived in the Wau area, in 1924, the first thing that struck me was the wealth of timber' (Booth 1939: 36). The government, following the advice of the inspector-general of forests, sought to license an 'established firm' as the forested area was deemed too small for individuals like Booth (*Pacific Islands Monthly* 1939: 27; see also Lane Poole 1935). While Booth had proposed water transport, the government at the time looked to the completion of a Wau–Salamaua road as a more realistic option. Ultimately, because of these problems, timber production continued to serve local needs in the years leading up to World War II.[4]

In extracting timber for the mining concerns, local practices and rights over such resources were quickly sidestepped. This was done through waste and vacant declarations, as well as devaluing the abilities and understandings of indigenous peoples. Taylour and Morley, in addition to describing the Melanesian population as 'poor specimens, constitutionally weak, backward, and without any form of religion' (1933: 40), also explained that they know 'nothing of rotation of crops', burning and cultivating until they have 'exhausted [the land's] capacity to further produce foodstuff' (ibid.: 23). The 'native' farmer is thus wasteful, but 'under the civilizing influence of the Administration, the Missions, and industrial organisations, the native is improving physically, mentally, and morally' (ibid.: 40–1). Thus, not only were miners justified in their sound use of timber and minerals, doing so would serve the local population as well. Benefiting from the model behaviour of the growing expatriate mining community, they reasoned that indigenous populations would learn to better manage their own resources by following European practices. This discourse of improvement becomes important when considering contemporary conceptualisations of community forestry below. But it also differs from the professional forestry discourses emerging simultaneously with early miners.

In contrast to the emphasis placed on industry use-values by miners and government officials in the mining office, government forestry officials emphasised the potential for an industry and the wastefulness of the mining community. In his 1935 survey of timber in the Morobe goldfields, Lane Poole was concerned with the value of trees and highly critical of the fact that the timber was not being used appropriately by mining companies. Cedar, for example, was used as an 'all purpose' resource when 'durability was necessary' (Lane Poole 1935: 5). After detailing the construction of a water race made completely of pine and cedar, Lane Poole complained, 'I quote this race in detail to show how far the *prostitution* of timber can go' (ibid., emphasis added). In Lane Poole's vision, different kinds timber have known purposes and the cedar deployed haphazardly in the goldfields would, in his view, be better put to use in the production of 'fine cabinetry' (ibid.). As the inspector-general of forests charged with making recommendations on the use of the resources found in the colony, Lane Poole enforced a set of values that sought to match grain and strength with products, fulfilling, in this case, the wood's finely tuned historical disposition for cedar cabinetry. Making races for mining from such a fine timber

---

4   At the time, Lae was not deemed a viable port, especially as Salamaua had been established as the coastal station supporting the development of the goldfields.

was simply offensive. Therefore, he argued that the mining community needed to better use the resources available to them. Miners, in his view, were not the 'civilizing influence' that they imagined themselves to be. They too were wasteful, knowing nothing of timber's value.

In contrast to claims by mining officials, it was not simply a matter of taking it out of local control and imposing colonial management. In another area of the goldfields, Lane Poole spoke favourably of Wahgi Valley forestry practices and the emphasis placed here is revealing:

> The destruction of the forests has resulted in a shortage of timber for hut-building and fuel; and *this has led the natives to adopt a forest policy*. Reports from the District and Assistant District Officers — Messrs. Melrose and Penglase — show that the natives in this region and in the Whagi [sic] Valley are making plantations use the Casuarina for the purpose. Timber is counted so valuable that the Government officers are obliged to pay for the tent poles and fuel required for their camps. These primitive people have, through the spur of want, developed a forest policy which shows them to be more fore-sighted then the average of their white brothers in Australia. (Lane Poole 1935: 4, emphasis added)

While Papua New Guineans certainly did not differentiate woods suitable for fine cabinetry, they did manage scarce resources and develop sustainable forestry practices. In contrast, the miners' wasteful 'prostitution of timber' was akin to the destruction of forests by swidden agriculture (which Lane Poole also derided). In both cases, the true values were not maximised.

As Wood (2005) notes, Lane Poole offers a more complex image of the colonial than the caricatures of state discipline described by Sivaramakrishnan. But more importantly for this chapter, it shows discrepancies in colonial authority and a diversity of ways in which forests were deemed useful by the colonial government. It also opens up an intellectual space in which local management is not only viable, but perhaps better than the expatriate mining community's use of the same.

In highlighting what must have seemed an irony from the perspective of colonial policy, Lane Poole also makes a series of recommendations for the development of an industry in the Wau–Bulolo area, which eventually comes to pass as practice. Aside from demanding that a route to the coast be identified, the establishment of regional offices and the training of technical experts, he strongly emphasised the need to develop a market internal to Papua New Guinea such that the colony would be self-sufficient in the production and maintenance of its own timber resources. Community forestry efforts in the village of Elauru reflect this call.

Logging Biangai Landscapes

For Biangai, trees are an important part of the social and physical landscape in which they work and live. The forests are home to animals, plants and spirits that are important for sustenance, shelter and cosmology. Hunting paths cut through the forests signify family connections, and stands of economically important fruit and nut trees (especially *Pandanus julianettii* and *Pandanus conoideus*) hold named relationships with specific individuals. In mourning songs (or *yongo ingi*), the soul of the dead is imagined as travelling through forested paths, where specific rivers and trees are evoked to tie the deceased to a lived landscape (Halvaksz 2003). It was perhaps in the construction of a men's house where the relationships between trees, families and people were most clearly displayed. Each family was responsible for providing a single post; one for each corner, the centre walls, a central post and ridge beam. Through its construction, the placement of trees in the men's house marked social relations rooted in land, connecting named kinship groups (*solonarik*) to each other and to the land from which the building materials were drawn. I argue below that trees made into timber have the potential to be both links in social networks as well as marketable products. But this also requires that we view the market more broadly as an essential network.

Community forestry was largely introduced into Papua New Guinea by missions and coastal plantations (Martin 1997). However, as detailed above, the Wau–Bulolo Valley has a unique relationship to such practices as they emerged in conjunction with gold mining. In the 1920s, small-scale timber operations initially supplemented the profits of both individual miners and the operations of the larger groups that eventually came to dominate the valley (Healy 1967; Sinclair 1998). The industry took off as the two larger operations, Bulolo Gold Dredging and New Guinea Goldfields, developed profitable timber subsidiaries that were key industries in the post–World War II mining economy.[5]

At first, companies acquired timber rights on government lands around the townships of Wau and Bulolo. Once they had expended the timber in those areas, New Guinea Goldfields began to negotiate with the neighbouring villages, including Biangai. Attempts to access stands closer to Wau were made in 1967–68, with New Guinea Goldfields seeking permission from the Department of Forests for timber rights on two areas of 'native owned land' (Lae Archive 1963/68: 26). The communities, it was emphasised in the supporting correspondence from District Commissioner H.P. Seale, were 'very keen to

---

5   Bulolo Gold Dredging eventually established a plywood factory in 1952. Surrounded by klinkii and hoop pines, Bulolo would become the centre of forestry and forestry research with the establishment of the Bulolo Forestry College. To this day, Bulolo continues to be a centre of forestry activities.

go ahead' (ibid.: 184). However, after an investigation by a forest officer, the department director expressed concern that 'the timber is very scattered, that the locals are more interested in an access road rather than a large cash payment, and that considerable work will be required of one of [the government's] officers to determine ownership' (ibid.: 63). Similar motives can be seen in subsequent logging deals discussed below and they reflect generally different goals and desires for partaking in timber schemes.

It was under the newly passed *Forestry Act 1973* that Elauru, Winima and Werewere, the three villages furthest from the town of Wau, became participants in the timber industry (Mitio 1981, 1984). The *Forestry Act (Amalgamated)1973* provided two processes for gaining access to timber. The first allowed for the declaration of a Native Timber Authority, a licence that allowed direct dealings between companies and communities over small areas of land. Under such schemes, logging was allowed only in so far as it fulfilled the construction needs of the company. In the second, the state acted as a broker in negotiating a Timber Rights Purchase agreement on behalf of communities. Elauru first contracted with New Guinea Goldfields under the smaller Native Timber Authority. But eventually a Timber Rights Purchase agreement was also negotiated on their land. Their experiences with the two arrangements contrast greatly.

In their version of this history, Biangai initiated the contact, as they were eager for the sorts of benefits accrued by the Biangai communities closer to the town of Wau. An elder described the events as follows:

> The story goes like this. A road just doesn't appear out of nothing, but it comes from the efforts of this village, Elauru. This land was here and the company sent word that it would like the trees in our forests. A good number of us, all papagraun [landowners], went to the NGG [New Guinea Goldfields] office and said to them, 'Why are you looking for trees? We have trees.' They asked, 'Where?' We told them in Elauru, and they wanted to know what kinds. We said, bush trees and a kind of pine tree, and the next day they came to look. (interview with Elauru elder, 2002)

As Elauru villagers tell the story, they were eager to have their land logged as it meant the extension of the road to their village, facilitating access and transport between their homes and the town of Wau. It was development as they understood it. But a road is more than just a path through the valley. Like hunting paths followed by men, or the winding routes that they take to their gardens, roads enfold histories, social relations and stories. Roads shared by family bind that family together, and one might expect that a logging road implies more than a practical relationship to the space. Such relations, and the desire to establish and maintain them, are not captured in the director of forests' concern that Biangai are more interested in roads than cash payments. In fact, roads have more meaning. In his narrative, the Elauru elder is staking a claim on the relationships that the road makes possible. In this account of the coming of the road, these sentiments are apparent:

R: They carried bows and arrows, two women in front, pulling their bows tight. Other women came behind singing, behind the two in front. They came all the way up to this point.

Jamon: The road crews were there too?

G: They were at the ceremonial gate below, passing through with us.

K: All of the men and women were performing singsings, pulling them up the hill, and a bulldozer pulled a tree behind the women.

G: Now they came up, passing all of us up to the edge of the trees.

Jamon: This was the first tree they had removed from the ground?

All: Yes

[they restate some of the details again]

Jamon: Now, the women were pulling their bows?

K: Yeah.

Jamon: Why?

K: Oh man … [starts to laugh] … because …

R: They didn't want to kill them!

K: They wanted to kill the whites, or maybe the driver. [laughing, others join in]

K: [still laughing] To show them that it is enough that one of them would get this arrow here [points to chest], but … it's ok

G: We said, 'Now, God's word has come, the church is here, and it is ok … take our trees and go'.

K: So, they took half from one area. Finished with that, they took half from places where we hardly ever go. That's ok, we said, you can take them and go. And us, as angry as we should be, enough that one of you took this arrow too, but … [laughs]

The trees were not simply disposable, and their removal was a significant event. But the road was marked as well through the performance of leading the bulldozer along the newly created path; a parade of sorts celebrating the connection between company, communities and the town of Wau.[6] That they might kill the company employees was funny, but the parade was marked by sadness (women remember crying on such occasions). The account reflects the sort of playful, yet real, tension that is demonstrated in Biangai debates about land and resources; one that is highlighted in their relationship with past and present development partners. Songs were also composed for the occasion.

---

6   The practice of pulling guests into a meeting is common. On numerous occasions, I witnessed, or was myself the subject of, such performances.

These were songs of sorrow, or *yongo ingi* (literally, women's songs), for the loss which the women sang as they tightened the strings of their arrows. *Yongo ingi* mark the passing of a loved one, noting their emplacement on the landscape, where they came from, where they will journey too; revealing relationships among the living and the dead, human and non-human (Halvaksz 2003). The songs composed for this occasion marked a similar passing. And as with the death of a family member who died too soon, where sorcery and mischief were suspected, they prepared their bows and arrows for the occasion. Removing the trees evoked connections with ancestors, ones that were forgone for the sake of a new relationship with the company, and through the company-created road, to the town of Wau.

But the final detail, the piecemeal extraction of timber from different locations, is important to the valuation of the trees as well and the extension of social relationships beyond the ancestral ones. Timber removed from different locations distributed compensation and established ties between specific Elauru landowners and New Guinea Goldfields. When evaluating the different regimes that have sought Biangai resources over the years (mining, logging, conservation, cash cropping), Biangai always spoke highly of logging. Many favoured it because of the distribution of benefits, as well as the enhanced communal social relations established with the representatives of the company. Contemporary gold mining concentrates on a single location, focusing the network on a smaller group, while logging can activate a much more expansive group of kin as timber is dispersed across the landscape. Biangai preferred the latter.

Under the Native Timber Authority, villagers were actively involved in 'witnessing' the selection and cutting. Individual landowners participated in identifying areas and measuring logs, and were paid according to an agreed price per super foot (see also Mitio 1981). Disputes were limited mostly to families, which is not to say that they were not substantial. Compensation money from logging was fought over, but more often than not redistributed somewhat fairly and with a great deal of investment in projects that spoke more of a commitment to relationships than to self-interests. For example, Elauru villagers spent a large part of their compensation on the construction of a permanent church building, using store-bought materials and employing a carpenter from the coast. Additionally, families receiving compensation payments used some of the money to cover their ancestors' graves with multi-tiered cement monuments. When I asked why this was done, it was said that these ancestors also hold title to the forests, and 'you have to treat them well'.

In contrast, Timber Rights Purchase agreements resulted in much broader disputes, both among families and between villages. Mitio (1981, 1984) analyses the disputes that eventuated between Elauru and Werewere as the forest-covered hills between the two villages were declared a Timber Rights Purchase area by the government. His discussion emphasises the lack of close contact between resource owners and the daily practices of the logging companies. Mitio concluded that these disputes resulted from colonial assumptions about the organisation of all Papua New Guinea communities into unilineal clans, when in fact the Biangai are cognatic, granting both men and women rights in resources. Individuals can hold multiple titles in different named resource groups called *solonarik*, making the assignment of individuals to clans a futile process. Yet, this was done at the beginning of the Timber Rights Purchase agreement, and continues to inform (or confuse) debates around mineral compensation today. As royalty payments were made to 'officially' declared clan leaders without clear identification of which specific timber resources were in question, those with secondary, or even primary, rights in a specific tree might never be compensated and, more importantly, they might not be acknowledged as part of the benefiting group.[7]

The contrast between these two approaches to timber extraction highlights three important factors for contemporary community-based timber production. First, in addition to the short-term financial gains, through participation and employment in the Native Timber Authority, villagers gained experience in forestry that remains useful. As discussed below, it was through employment that individuals learned skills that have ensured the longevity of community-based timber production. Second, smaller projects that provided direct returns to specific right holders were less likely to lead to disputes and more likely to result in some sort of locally meaningful investment. Third, and perhaps most significantly, the Native Timber Authority allowed for a greater sense of participation and control over the process, if not the actual practices. Subsequently, community members evaluated this experience more positively.

The employment of Biangai men during the Native Timber Authority agreement increased their sense of control as they ensured a greater degree of transparency through participation in the logging and processing of their resources. One of those employees, K–, benefited greatly from working with New Guinea Goldfields. At first hired for manual labour, he was eventually trained in the operation of much of the equipment, tripling his income in the process. At one point, he was given his own crew and a company certificate in chainsaw operations. While he described working with the chainsaw as 'good work', he complained that if he went to work for a company, 'it would be like living in a prison'. Unable to take

---

7   The disputes between villages follows a similar logic, and continued during my research in 2002 as a group of Werewere villagers tried to establish a cattle project on the area that was cleared by logging.

breaks when he wanted, or work on different parts of the job, he would feel trapped, lacking control over his own life. Perhaps this sentiment is a factor in shaping how community forestry is practised today.[8]

The Kuper Range Wildlife Management Area

The Kuper Range Wildlife Management Area began in 1989 after discussions between Wau Ecology Institute (a local non-governmental organisation (NGO)) and Elauru village. Identified as part of a priority area for biodiversity conservation (Beehler 1993; Sekhran and Miller 1995), and coupled with the Wau Ecology Institute's desire to move its overrun Kaindi research station, the Kuper Range was established in 1992. Initially 500 hectares of land was offered by a group from Elauru.[9] The use of this facility by tourists, researchers, students of the Bulolo Forestry College and a National Geographic film crew increased village interest in the project and more families contributed land. Throughout, the conservation area was couched in terms of it meeting the expectations of development (see West 2006 for similar development discourse). The Wau Ecology Institute encouraged such thinking, funding numerous 'sustainable' development projects.[10] Along with beekeeping, a pig farm, handicrafts, bottle recycling, a small trade store and ecotourism, a small-scale community forestry project was included for K– and his family. Wau Ecology Institute used K3,000 to purchase a chainsaw, tools and a limited supply of fuel and oil. In addition, training was promised in ecological methods and equipment maintenance. While the training never eventuated, K– and others possessed enough experience and ingenuity to continue a community-based forestry project. Of the many projects funded, it is the only one that was still in operation at some level in 2005.

Early on, using chainsaw milling, K– was able to bid on and gain a number of contracts in and around Wau, in addition to casual labour for Wau Ecology Institute. The field station manager, who was charged with maintaining the paths around the proposed conservation area, was given a second chainsaw. But as discussed below, it too became significant in the local village-based

---

8   Seeking a change in his life, K– applied to become an evangelist, and was accepted by the Lutheran Church's school in Finschaffen for training. For some 10 years timber did not play a major role in the lives of Elauru villagers. Shortly after K– left for training, New Guinea Goldfields closed its sawmill in 1976, having failed to comply with stipulations of the Forestry Act (Sinclair 1998: 442). The company did not have a formal development plan for managing the forest areas under its control. Its contract in Elauru was left unfinished. When the idea emerged to start an eco-timber project in association with a small-scale Wildlife Management Area, K– was ready to participate.

9   Gold prospecting and squatters at Kaindi made the area less useful for biological research into Papua New Guinea's unique flora and fauna.

10   The Wau Ecology Institute received a grant from the New Zealand High Commission to fund a landowner incentive project. The institute stated two objectives in this grant: (1) to assist Elauru villagers in officially registering the Wildlife Management Area and (2) to establish an incentive program that would 'compel people not to sell their forests for commercial logging' (Wau Ecology Institute n.d.: 10).

economy. With the exception of the field station and ecotourism, none of the projects proved to be year-round successes. Many are long forgotten, and even ecotourism, once subject to the temperamental relationship between the NGO and the community, ended in 2003. Their relationship deteriorated as villagers increasingly questioned Wau Ecology Institute's expenditures, and NGO staff became annoyed with weekly requests from Elauru villagers for compensation, development and/or support.

Throughout, K– remains a visionary, constantly planning ways to improve his project. He imagines opportunities, and programs that will empower not only himself, but other villagers as well. He spoke to me often about how his project would expand with consistent access to transport and better equipment. But most notable was his vision of a certification program that would allow him to train and certify other Biangai in the use of the chainsaw. He would teach others the skills necessary for participation in the community-based timber production project and employment opportunities in the local labour market.[11] In the meantime, his success has been limited to bidding on local contracts and acquiring occasional work removing trees that threaten to fall on village homes, or using crude chainsaw milling methods to cut lumber for local use.

Biangai Forestry and the Community

The lack of economic success and full-time commitment to the community forestry project ensures that it never meets the expectations of international aid agencies and NGOs. Many would not even deem this a community forestry effort. Elauru's community effort neither grows, nor increases profits from year to year. Instead, the project continues to make meagre economic returns, without ever really completely falling apart. In part, this results from the retention of skills and equipment necessary to maintain the project. Furthermore, while much of the area has been logged, the timber resources are not entirely depleted, as with some areas in Papua New Guinea. But more importantly, the local demand for sawn timber remains, even if only on an intermittent basis. This is demonstrated in the village housing market.

Houses in Biangai villages are always under construction. Bundles of bamboo, posts symmetrically embedded in the ground, and skeletal walls temporarily braced against the wind can always be found in some corner of the community of Elauru as families work at replacing old deteriorating structures. They are not company built, but built with the help of many kin, and this construction

---

11   He also imagines his own 'integrated conservation and development project', one that is full of contradictions. On one hand, he wants to recreate the facilities of Wau Ecology Institute on his family's land, cutting the timber needed and constructing offices, a hostel, laboratories and a zoo on his own. However, he would do so using timber from the very forests that serve as the centrepiece of the conservation area.

process adds meaning to the structure. When I helped hammer or carry a post or supplies, the homeowner would laugh and say that they will always think of me when they stand near the nail that I had just pounded in. Such comments were also applied to houses that were finished, as they reminded the residents of kin who had passed away, or moved to distant towns. My adoptive mother in Elauru often recalled how the walls of their house (made of hand-shaped planks fitted like siding) were originally used in the 1950s by her father. The flooring of sawn timber was from the house of my adoptive father's father, made from New Guinea Goldfields' staff houses built during the initial logging operations. His house was torn down at the end of a period of mourning a year after his death, and the usable components now circulate in the homes of his children as reminders. The timber, cut by New Guinea Goldfields from plots of land connected to the family, becomes part of an extended memory, folding the tall pines of the past landscapes, homes and lives into present buildings and contemporary social relations.

Houses are often recycled in this way. During research in 2000–02, Winima village was in the process of moving as the government had deemed it in danger of being destroyed by landslides. Instead of building entirely new structures, each building was almost entirely recycled into the new location. Houses and house parts travelled along the road, with only the most significantly worn pieces replaced by new bush materials. Such recycling serves as a memory of past labour and past labourers. The destruction of an old home can be quite emotional, especially if it is the home of one who has recently died, or was built with the help of such persons. A post bares both the connection to the land and the 'hand mark' of human labour.

The typical house embodies the work of an entire extended family, and the forests that they command. Construction begins with the cutting of hardwood posts, often designated for such purposes a year or more in advance. Cut marks made across the base of a tree confirm its eventual purpose. After posts are designated, the trees are felled and shaped by hand with axe and machete, and left to dry. A single post could take a half-day to shape. The frame of the house is a similarly intensive production, accumulated over the course of weeks if not months. Only the bamboo and *kunai* are assembled in the final moments, as neither can sit too long and remain useful. Their assemblage indicates the final stages, with the threading of grasses for the roof, and splitting and weaving bamboo floors and walls. Often, such work might entail an entire community. Forestry practices facilitated this work, and were valued for making homes and wood products that would outlast their predecessors. Y–'s house offers an example of such practices.

Prominently situated at the entrance to the village, Y—'s house was slow to take shape. For months, 16 iron posts stood alone in his area. These were purchased at a considerable discount through a relative who was the headmaster at the vocational school in Wau where students are trained to make such products. Many watched with excitement and amusement as Y— and a teacher from the elementary school carefully marked the area, and checked that the posts stood perfectly upright before they poured in the cement. But then, nothing else happened. More than six months passed, and the posts remained unattended.

One morning while most of the community was preparing to spend the day in their gardens, there was much movement around Y—'s forever-emerging home. Towards the end of the coffee season, he had purchased fuel and oil for one of the two village chainsaws and work had long been under way in a small crop of trees on his land. Using this fuel, the operator would be able to cut timber for both Y— and himself. The delay came from the difficulty that the chainsaw operator was having with fashioning the many sizes and transporting the finished products from the bush to the village. They managed to cut the thicker cross beams for the floor, walls and roofs, but the thinner sizes kept splintering and breaking under the crude method of chainsaw milling. With the materials finally carried to the village, Y—'s floor started to take shape.

Some months passed, and Y— and his extended family had finally raised enough money through the sale of coffee to purchase the remaining timber from a neighbouring village sawmill in Wandumi and from an expatriate in Wau who purchases stands of timber to operate his portable sawmill. The timber and its transport cost money, but Y— viewed the expenditure as worthwhile. Other villagers were equally impressed, as only three other homes were made of sawn timber, most of which was taken from logging camp scraps. The walls of his house finally started to take shape.

Y— told me, 'everyone has already said, you are building a real house ... just like those in town ... not at all like a village house'. But this was far from the complete story. While it looked much like the homes in the town of Wau, he had not constructed his house in the same way as a 'town' house. Instead, using his limited funds, he had mobilised his network of kin and other relations around this effort. Much of the money for timber and supplies, as well as the cups of tea, plates of rice and canned meat to feed those who worked a few hours raising his walls and roof, came from his extended family.

While the expatriate mill did not cut him a deal, his family in Wandumi helped by saving 'flawed' pieces for him to purchase at a discount. A relative employed by the Works Department in Lae, and another who was the headmaster of Wau's vocational school, both supplied him with money and 'excess' supplies from their stores. Many in the community (myself included) hammered and sawed, donating labour and tools towards the effort. Most of the newly constructed houses in town were built using prefabricated forms, purchased in Bulolo and assembled by local carpenters, if not by their future residents. Y– could not afford such expenses, but he could utilise his network of kin and the availability of skilled community foresters in his efforts. Many now plan to build similar homes, suggesting an increase in local demand for timber products. With the opening of gold mining at nearby Hidden Valley, it was likely that increased cash flows would facilitate this effort.

Conclusion: Determining Success in Community Forestry

In his review of community forestry in Papua New Guinea, Martin noted the collaborative potential for such efforts, as the influx of 'small-scale forestry, indigenously owned and managed, offers a theoretical potential for diverse interest groups to realise their objectives in cooperation with each other' (1997: 270). These groups seek to address ecological, economic and political concerns providing sustainable alternatives to large-scale resource extraction (such as described in Filer with Sekhran 1998). Others have concluded that such projects contribute positively to community and environmental sustainability (Salafsky et al. 1997/8). However, the systematic lack of training and reliable transport, over-reliance on local markets, internal friction, and project sustainability continues to hinder community-based timber production. Furthermore, Martin is critical of part-time efforts that fail to meet commercial standards as communities are torn between social obligations (such as funerary expenses and feasts, subsistence and cash-crop gardens, and village politics) and full participation in community forestry activities. From the perspective of sustaining a capitalist system of timber production, full-time participation is necessary for project success.

In contrast, Gilmour and Fisher (1997: 28) argued that regardless of economic motivations, community forestry 'is essentially about access to and control of forest resources' (see also Peluso 1992). This suggests a slightly different point of analysis; one that focuses on local empowerment, where political and social sustainability are privileged over economic profitability. As Bun and Baput (2006) point out, community forestry is a process and is different from industrial logging where quick profits are the goal. Here, I have argued that while Biangai communities struggle to succeed according to a formal economic model, they do succeed in terms of providing villages with alternative practices and opportunities that they might otherwise not enjoy. Such successes raise questions about development models, whether they are labelled sustainable or not. Instead of focusing on profit, export markets and the imposition of a strictly capitalist model, I argue that such projects help extend the base of community resources, allowing for innovative and creative engagement within a transitional economy (Gudeman 2001). Moreover, we must attend to the different valuations of trees as timber and trees as embedded in community networks (Leach 2001). As objects that 'circulate', trees and their parts become differently entangled in systems of meaning that simultaneously facilitate both Biangai and corporate values. Values are at play when trees transact relationships among Biangai. And they are at play when Biangai, trees and their products engage in wider networks of exchange. Trees for Biangai are not simply utilitarian items of market value. Even when used in the building of homes, the value of trees is as much about the relationships they evoke as the qualities of the species themselves.

More broadly, Gudeman (2001: 158) has argued that development results not from capital accumulation 'but from innovation in the relationships of society'. Such innovations result not merely from capital investment, technical transfer and education, but even in an age of globalisation innovation emerges from social networks. House construction is not a novel development for community-based timber projects (Filer with Sekhran 1998: 299). Critiques of such practices emphasise the diversion of labour from more profitable endeavours and towards providing for local needs. However, I argue that as they provide opportunities for local innovation, the development of a specialised rural labour and improved living standards with more permanent structures, such projects are successful at extending the economic base of communities, and thus future opportunities. In many ways, contemporary Biangai practices follow the model of early goldminers, where forest resources are affordably accessed so that effort can be given to other development efforts. During the time under consideration, coffee, conservation and gold offered other 'development-like' benefits. Coupled with forestry, Biangai can be seen as creatively engaging diverse economic opportunities.

New Guinea Goldfields has already extracted the most profitable tracts and a nearby company, PNG Forest Products, continues to fill the broader regional demand for milled timber, but at prices beyond the local village. The timber demands of the Biangai communities around Wau are intermittent, based on seasonal cash flows from coffee sales, and a limited, but growing, interest; particularly as the Biangai are attracted to all things modern. Thus, it remains important that they have local resources to meet their needs (as Lane Poole once suggested). The community forestry projects in Elauru, as well as neighbouring Wandumi, seem to be doing just that.

While not meeting NGO and capitalist expectations for growth, Elauru's community forestry efforts suggest much more about community sustainability, focusing attention on developing an economic base and community independence in contrast to a reliance on fluctuating markets and dependence on the external advising of temperamental NGOs. K–'s many plans also show the promise of this effort, one that is not captured in the analysis and critique of similar projects. As NGOs and international donor agencies rush to fund community-based timber production, it is import to acknowledge that there are different measures of success.

References

Beehler, B. (ed.), 1993. *Papua New Guinea Conservation Needs Assessment* (Volume 2). Washington (DC): Biodiversity Support Program and Boroko: Papua New Guinea Department of Environment and Conservation.

Booth, C., 1939. 'New Guinea's Timber Wealth: A Pioneer's Experience at the Hands of Bureaucracy.' *Pacific Islands Monthly*, 16 May.

Booth, D., 1929. *Mountains, Gold and Cannibals*. London: Morrison and Gibb.

Bun, Y.A. and B. Baput, 2006. 'Community Forestry Benefits Customary Landowners: Case Study on Madang Province Papua New Guinea.' ITTO, Forest Trends, Rights and Resources Initiative (CFE Case Study Report).

Deleuze, G. and F. Guattari, 1987. *A Thousand Plateaus: Capitalism and Schizophrenia*. Minneapolis: University of Minnesota Press.

Filer, C., with N. Sekhran, 1998. *Loggers, Donors and Resource Owners*. London: International Institute for Environment and Development in association with the PNG National Research Institute (*Policy That Works for Forests and People*, Country Study 2).

Gilmour, D.A. and R.J. Fisher, 1997. 'Evolution in Community Forestry: Contesting Forest Resources.' In M. Victor, C. Lang and J. Bornemeier (eds.), *Community Forestry at a Crossroads: Reflections and Future Directions in the Development of Community Forestry*. Proceedings of an International Seminar held in Bangkok, Thailand, 17–19 July 1997 (RECOFTC Report 16).

Gudeman, S., 2001. *The Anthropology of Economy: Community, Market, and Culture*. Malden (MA): Blackwell Publishing.

Halvaksz, J.A., 2003. 'Singing about the Land among the Biangai.' *Oceania* 7(3): 153–69.

Halvaksz, J.A., 2006. 'Cannibalistic Imaginaries: Mining the Natural and Social Body in Papua New Guinea.' *The Contemporary Pacific* 18(2): 335–59.

Halvaksz, J.A., 2008. 'Whose Mine Closure?: Appearances, Temporality and Mineral Extraction along the Upper Bulolo River, Papua New Guinea.' *Journal of the Royal Anthropological Institute* 14(1): 21–37.

Healy, A.M., 1967. *Bulolo: A History of the Development of the Bulolo Region, New Guinea*. Canberra: The Australian National University, New Guinea Research Unit (New Guinea Research Bulletin 15).

Jacka, J., 2001. 'Coca-Cola and *Kolo*: Land, Ancestors and Development.' *Anthropology Today* 17(4): 3–8.

Lae Archive, 1963/68. 'Forests Upper Watut also Nayti Forest Land.' Lae Archive, Box 263. Lae, Morobe Province, Papua New Guinea.

Lane Poole, C.E., 1935. 'Report on the Forests of the Goldfields of New Guinea: Together with Recommendations Regarding a Forest Policy for the Whole Territory.' Unpublished TS, 1935. NAA, A1 1938/802.

Leach, J., 2001. 'Land, Trees and History: Disputes Involving Boundaries and Identities in the Context of Development.' In L. Kalinoe and J. Leach (eds), *Rationales of Ownership: Ethnographic Studies of Transactions and Claims to Ownership in Contemporary Papua New Guinea*. New Delhi: UBS Publishers' Distributors and Port Moresby: UPNG Law Faculty Publication Unit.

Li, T.M., 2007. 'Practices of Assemblage and Community Forest Management.' *Economy and Society* 36(2): 263–93.

Martin, R., 1997. 'Small-Scale Community-Based Forestry: Issues in the Conservation of Papua New Guinea's Biodiversity.' In C. Filer (ed.), *The Political Economy of Forest Management in Papua New Guinea*. Boroko: The National Research Institute and The International Institute for Environment and Development (NRI Monograph 32).

Mitio, N., 1981. Biangai Marriage and Its Relationship to Kinship and Property: The Case of Werewere Village, Wau District. Port Moresby: University of Papua New Guinea (Honours thesis).

Mitio, N., 1984. *Factors Affecting People's Attitudes to Company Exploitation of Their Resources: The Case of Timber*. Boroko: Institute of Applied Social and Economic Research (Special Publication 9).

Pacific Islands Monthly, 1939. 'New Guinea's Inland Timbers: Permit for One Firm Only.' *Pacific Islands Monthly*, 15 March.

Peluso, N., 1992. *Rich Forests, Poor People: Resource Control and Resistance in Java*. Berkeley: University of California Press.

Salafsky, N., B. Cordes, M. Leighton, M. Henderson, W. Watt and R. Cherry, 1997/8. 'Chainsaws as a Tool for Conservation? A Comparison of Community-Based Timber Production Enterprises in Papua New Guinea and Indonesia.' London: Rural Development Forestry Network (Network Paper 22b).

Sekhran, N. and S. Miller (eds), 1995. *Papua New Guinea Country Study on Biological Diversity: A Report to the United Nations Environment Programme.* Port Moresby: Department of Environment and Conservation, and Africa Centre for Resources and Environment.

Sinclair, J., 1998. *Golden Gateway: Lae and the Province of Morobe.* Bathurst (Australia): Crawford House Publishing.

Sivaramakrishnan, K., 1995. 'Colonialism and Forestry in India: Imagining the Past in Present Politics.' *Comparative Studies in Society and History* 37(1): 3–40.

Taylour, H. and I.W. Morley, 1933. 'The Development of Gold Mining in Morobe, New Guinea.' *Proceedings of the Australasian Institute of Mining and Metallurgy* 89: 1–91.

Waterhouse, M., 2010. *Not a Poor Man's Field: The New Guinea Goldfields to 1942 — An Australian Colonial History.* Sydney: Halstead Press.

Wau Ecology Institute, n.d., 'Kuper Range Conservation Area and Incentive Promotion Project. Grant Proposal Submitted to the Government of New Zealand.' Unpublished grant application.

West, P., 2006. *Conservation Is Our Government Now: The Politics of Ecology in Papua New Guinea.* Durham (NC): Duke University Press.

Winkel, G., 2010. 'Foucault in the Forests — A Review of the Use of 'Foucauldian' Concepts in Forest Policy Analysis.' *Forest Policy and Economics* 16: 81–92.

Wood, M., 2005. 'Charles Lane Poole and Early Forest Surveys of Papua and New Guinea.' *Journal of Pacific History* 40(3): 289–309.

5. The Impact of Mining Development on Settlement Patterns, Firewood Availability and Forest Structure in Porgera

JERRY K. JACKA

'Wood for Sale': The Firewood Crisis in Porgera

In the late 1990s in the Porgera Valley, Porgerans frequently described a common difference to me between their country and mine — the United States. For Porgerans, the common refrain was that Papua New Guinea (PNG) was a free country (*fri kantri* in Tok Pisin). Whereas everything in the United States costs money, everything in PNG was free — firewood, water, food, building materials and so forth. All one had to do was walk into the surrounding rainforest, cut down trees for lumber, harvest lianas for lashing together timbers, build a house, make a fence, and then plant a garden in the cleared space for food. Water for crops and people would be provided by the abundant rains and numerous springs found throughout the landscape.

By 2006, however, things in Porgera were no longer quite so 'free'. *Selim diwai*, or 'wood for sale', signs lined the Enga Highway along the several-kilometre stretch from the security checkpoint at the Mt Maip boom gate to the outskirts of the world-class Porgera Gold Mine. Most surprisingly, the wood that was for sale was not pre-cut lumber for building, but was just plain firewood. Bundles of four 6-foot-long split pieces of firewood were being sold for K4 (about US$1.20) apiece. Not only was I shocked to see firewood being sold, but in the six years since I had last visited Porgera, I couldn't but help notice the intensive cutting of forest that had occurred along the highway as well. In fact, this is why I had returned to Porgera. One of the things I was interested in trying to understand was how migration into the valley was impacting forest resources. From a population of around 9,000 when the mine was first built in 1990, to an estimated population of over 40,000 today, I was interested in exploring how access to and control over resources, and the resources themselves, were faring in the era of mining development.

At the outset I want to stress that the relationship between forest exploitation and population increases cannot be explained in reference to simplistic associations related to carrying capacity and other kinds of 'balance of nature' arguments (Scoones 1999). Instead, I want to examine the issue of resource extraction and access to resources through a political ecology of development and human–forest dynamics that addresses political economy *and* ecology equally (Vayda and Walters 1999). As a consequence, I see the firewood crisis in Porgera as mutually constituted by sociocultural and ecological factors. From a sociocultural perspective, I explore Porgeran conceptions of land tenure, but do so from the recognition that different ethno-ecological zones related to relative altitude structure the claims that different actors can make regarding certain resources. From an ecological standpoint, I highlight forest succession dynamics in these different altitude zones and link them to economic and cultural uses by Porgerans to understand some relationships between forest exploitation, ecological succession, and sustainability of harvest in the different altitude zones.

Fuelwood Demands and Deforestation in Rural Global Settings

About half of all the timber cut globally is used as fuelwood and, of this amount, around 90 per cent is used in developing countries by rural populations for cooking and heating purposes (Parikka 2004). While the cutting of trees for fuelwood is sustainable in some cases, it has been estimated that wood cutting in tropical forests for lumber, charcoal production and fuelwood 'constitute 68% of the proximate causes of deforestation in Africa, 89% in Asia, and 51% in Latin America' (Naughton-Treves et al. 2007: 233). Even in areas where deforestation is not occurring, harvesting can result in changes in species composition and ecosystem function, as fast-growing secondary species replace slow-growing, old-growth species. Moreover, as is well documented by a number of ecologists, often species compositions after human disturbance do not return to levels that existed prior to human use (Holling 1973). Nevertheless, management plans in many countries wrongly assume that successional pathways in cleared tropical forests will lead to ecosystems similar to the ones that existed before clearing, that this will happen fairly rapidly, and that species richness will be similar to the original forest (Struhsaker 1997; Chapman and Chapman 2004). There are of course no forest management plans in place in Porgera, as land management is in the hands of customary social groups. However, as the data herein indicate, more holistic management plans and socio-economic impact studies need to take into account long-term ecological changes brought not just by the impact of whatever resource is being extracted, but also by the impacts of migrants, many who will stay in resource development zones for years, if not decades (Dambacher et al. 2007). In PNG, it is estimated that 23 per cent of the rural population lives in areas where firewood is sold, and between 1990 and 1995 firewood sales generated an estimated K5 million income annually for rural villagers (Bourke and Harwood 2009: 392). The expansion of development-related economic activities in the coming years will only increase the tensions surrounding fuelwood availability as land is cleared through logging, plantation development, and mining and petroleum ventures.

The protection of forests in PNG is critical as the island of New Guinea contains the third-largest extant bloc of rainforest in the world (Primack and Corlett 2005). New species are constantly being discovered and little ecological work has been done since the transition to independence in the mid-1970s. In 1972, approximately 29 million hectares of rainforest covered PNG, which constituted about 62 per cent of the country's land cover. By the end of the twentieth century, between 3 and 5 million hectares of this forest had been cut.

Currently in PNG, the varying impacts of the drivers behind deforestation and the rates of deforestation overall are the source of some controversy (Shearman et al. 2008, 2009, 2010; Filer et al. 2009; Filer 2010). As a result of the PNG government's central role in the UN-REDD program[1] (United Nations Collaborative Programme on Reducing Emissions from Deforestation and Forest Degradation in Developing Countries), Filer and colleagues (Filer et al. 2009; Filer 2010) argue that the PNG government has inflated the past, present and future rates of deforestation in order to obtain the greatest benefits from the REDD program. The source of the controversy stems from three different studies of deforestation in PNG, all of which appear to demonstrate varying rates of deforestation:

---

1   Technically, the PNG government is involved in REDD+, which goes beyond just deforestation and degradation and includes conservation, sustainable management and enhancement of forests. See Gabriel, Chapter 10, this volume, for more information about REDD+.

- the Remote Sensing and Land-Use Project (RSLUP) associated with Shearman and colleagues (Shearman et al. 2008, 2009)

- the Forest Inventory Mapping System (FIMS), which Filer and colleagues rely upon (Filer et al. 2009; Filer 2010)

- assessments by the Food and Agriculture Organization of the United Nations (FAO 2005, 2010), which are derived from the FIMS findings (Shearman et al. 2008: 117–8; Filer et al. 2009: 813), although the data between the latter two do not match one another (see Table 5.1)!

Central to each of these studies are slippery definitions of such terms as: 'deforestation', 'degradation', 'primary forest' and 'primary rainforest'. Moreover, the data these studies draw on are scattered throughout various sources (much of it inaccessible grey literature) and the methods for which land-cover classification was determined (in the FIMS and FAO datasets) are not comprehensively detailed.

Deforestation comes from many sources: logging operations, natural and anthropogenic fires, plantation development, subsistence agriculture and mining. In many instances, forest is converted to grasslands, scrub or agricultural land, which may persist for decades (Shearman et al. 2009: 379). Shearman and colleagues (2008, 2009) list commercial logging and subsistence agriculture as the two primary drivers in the loss of primary rainforest, and attempt to demonstrate that increasing population results in loss of primary forests. This contradicts what shifting cultivators themselves say — who prefer to use secondary forests — and recent studies looking at agricultural intensification in PNG (Allen et al. 2001; McAlpine and Freyne 2001). I will address this in greater detail later in the chapter as the Porgera data support the intensification argument to some degree.

At issue in the deforestation controversy is how much primary rainforest existed in 1972 (the date at which the Australian government compiled most of the aerial photos of the country, which were subsequently used in making 1:100,000 topographic maps with land-cover features; both FIMS and RSLUP used this data for their 1972 baseline). According to FIMS, there were 29,317,500 hectares (ha) of primary rainforest in PNG in 1972. RSLUP figures state that there were actually 33,227,590 ha. As Filer et al. (2009) exclaim, this is a difference of almost 40,000 km^2! Given this discrepancy between initial numbers, RSLUP claims that nearly one-quarter (23.8 per cent) of PNG's forests were cut between 1972 and 2002, while FIMS reports a loss of only 10.9 per cent between 1972 and 1996. The source of this discrepancy perhaps lies in a slippage of how terms like 'intact forest', 'primary forest' and 'primary rainforest' are used by the RSLUP team and how 'total gross forest' and 'adjusted gross forest' were used by FIMS (see Filer et al. 2009: 813–4). The RSLUP team writes, 'In 2002 there were 28,251,967 ha of rain forest … with an additional 3,409,018 ha of swamp forest, 574,876 ha of mangroves, and 750,309 ha of dry evergreen forest. Of the rain forest, 25,332,253 ha were primary forest and 2,919,714 ha were secondary forest' (Shearman et al. 2009: 383). In summary, there are multiple classes of forest: primary rainforest, secondary rainforest, dry evergreen forest, swamp forest and mangroves. For 1972, however, the land-cover classes were 'intact forest, mangroves, scrub, grassland, and water. For the change analysis, land-cover classes were combined into two categories: forest (rain forest) and nonforest (non-rain forest)' (ibid.: 381). This raises the critical question of what happened to the 4,159,327 ha of swamp and dry evergreen forest from 2002 — were they included in the 1972 'intact forest' category? For the remainder of this chapter I will assume so, and for the purposes of clarifying and comparing the data will assume that these two 2002 land-cover categories were similarly sized in 1972.[2] This also helps to account for the 40,000 km² discrepancy between the RSLUP and FIMS data. Table 5.1 summarises these new amounts, removing 4,159,327 ha from RSLUP's 1972 estimate of over 33 million hectares of primary rainforest in 1972.

Table 5.1 Summary of FIMS, RSLUP and FAO studies[a] of deforestation in Papua New Guinea.

Study	Area of primary rainforest[b] (ha)	Area of deforestation (ha)	Area of degradation (ha)	Annual rate of deforestation (%)	Annual rate of degradation (%)	Total annual rate (%)
1972 FIMS	29,317,500	3,209,600	1,922,300	0.46	0.27	0.73
1996 FIMS	26,107,900					
1972 RSLUP	29,068,263	3,736,010	2,919,714	0.43	0.33	0.76
2002 RSLUP	25,332,253					
1990 FAO	31,329,000	2,985,000	1,886,000	0.48	0.30	0.78
2005 FAO	28,344,000					

[a] FIMS = Forest Inventory Mapping System, RSLUP = Remote Sensing and Land-Use Project, FAO = Food and Agriculture Organization of the United Nations.

[b] This is equivalent to what FIMS called 'adjusted gross forest area' (see Filer et al. 2009). FIMS data are reported in square kilometres; RSLUP and FAO data are in hectares.

Source: Shearman et al. (2008, 2009), Filer et al. (2009), Filer (2010), FAO (2005, 2010).

---

2   Of course they were not the same area, but it is probably safe to assume that there were no commercial logging operations in these forest types that would have diminished them from their 1972 area.

As Table 5.1 demonstrates, once the confusion generated by RSLUP's 1972 'intact forest' and 2002 'primary rain forest' is accounted for, the FIMS and RSLUP data are almost identical in terms of primary rainforest in 1972 and subsequent deforestation. Moreover, the total annual rate of deforestation and degradation are very similar across the three studies (see the last column in Table 5.1). Degradation in Table 5.1 refers to the 2.9 million ha of secondary rainforest as documented by the RSLUP group. This was forest that had been logged and was naturally regenerating. According to FIMS, this amount was 1.9 million ha; however, since PNG's logging industry was booming in the late 1990s (Filer et al. 2009, Figure 1; Shearman et al. 2009, Figure 5) this may account for this difference. The FAO figure of 1,886,000 ha of degraded forest comes from the sum of their categories of 'naturally regenerated forests' and 'planted forests' between 1990 and 2005. In Table 5.1, annual rates of deforestation and degradation have been linearly extrapolated. As argued by Shearman et al. (2009: 388), this ignores the recent increase in deforestation and degradation stemming from the uptake in commercial logging. As such, they modelled deforestation and degradation to each year based on the volume of timber exported, oil palm exports, increases in population and the intensity of El Niño events to account for logging, plantations, subsistence agriculture and fire as primary drivers (ibid.: 383). As such, they came up with annual rates of deforestation varying from 0.4 per cent in 1972–73 to 1.4 per cent in 2001–02, with a peak of 1.8 per cent in 1997–98. In primary forests accessible to commercial loggers, they calculated the rate of deforestation at 2.6 per cent per annum. These numbers are decried by Filer (2010: 151) as exaggerations stemming from the 'moral hazards' that REDD poses to 'rainforest nations' that seek to claim benefits from reducing greenhouse gas emissions from fictitious levels of deforestation and degradation. This is a point well made; however, recent data from the FAO show a rate of deforestation in PNG of 1.51 per cent per annum from the years 2005 to 2010 (FAO 2010). Importantly, as Filer et al. (2009: 813) note, a significant part of what is being classified as forest loss is part of the subsistence cycle of shifting cultivation in which more than 80 per cent of Papua New Guineans are engaged. However, in the absence of detailed ecological studies it is difficult to state with any certainty how an increasing population in PNG with larger demands on the forest will influence future forest regeneration, although the discussion later in this chapter deals with some possible implications based on the data in Porgera.

What I have tried to do in this section is sift through the data on deforestation in PNG and attempt to show that a recent controversy over the rate of deforestation isn't perhaps quite so controversial after all once confusion over terminology and land-cover categories is accounted for. Nevertheless, whether the annual rate of deforestation of primary rainforest is 0.46 per cent, 1.4 per cent or 1.51 per cent, these are still alarming numbers given the vast area of rainforest cut each year in PNG. Between 2005 and 2010, the FAO estimates that 427,000 ha of primary rainforest was cut each year in PNG (FAO 2010). This amount of deforestation currently ranks as the second highest in the world. Only Brazil cut more rainforest (more than five times as much as PNG, but Brazil has a forested area 18 times larger than PNG's rainforests).

At this juncture, in discussing forests, I don't want to lose sight of the trees for the forest, for as I discuss in the following sections, it is not the overall clearance of forest that needs to be considered in addressing the firewood crisis in the Porgera Valley, but rather where forest resources are being extracted as a function of altitude, successional dynamics, forest type and resource use rights.

Study Area – Ecological Characteristics

The Porgera Valley watershed (553 km²) is located in western Enga Province, PNG (5° 16' – 5° 33' S, 143° 0' – 143° 16' E). Elevations range from 900 metres in the northern part of the valley and climb to over 3,900 metres in the south. In 2002, based on satellite imagery, approximately 74.5 per cent of the watershed was covered in primary rainforest. Mean annual rainfall is 3,667 millimetres, with very little seasonality. In 30 years of rainfall data, there have been only three months that received less than 100 millimetres of rain. The mean daily minimum temperature is 11.8 °C, and the mean daily maximum temperature is 20.2 °C.[3] The remainder of the land cover in the watershed is grasslands (3.1 per cent), gardens (10.6 per cent), secondary forest (9.1 per cent) and mining/built up areas (2.6 per cent).

Ecologists and local groups alike offer multiple ways of defining the various forest zones in PNG. Robbins (1961) characterises forests from 3,000 to 9,000 feet (914–2,743 metres) as lower montane rainforest; similarly, Paijmans (1976) sets lower montane rainforest as occurring between 1,000 and 3,000 metres, whereas Johns (1982) defines lower montane as occurring at 300–1,000 metres in its lower limits and 1,500–2,000 metres in its upper limits, and mid-montane rainforest occurring at 1,500–2,000 metres in its lower limits and 2,700–3,000 metres as the upper limits. While many ecologists have tried to use features of forest structure (such as canopy height, presence/absence of buttressed trees and so forth) to delineate forest types, often these designations are characterised by linking dominant species to the divisions, such that lower montane is often associated with oaks (*Castanopsis acuminatissima*) and mid-montane with southern

---

3   Rainfall and temperature data come from the Porgera Joint Venture mine records. All readings were taken at Anawe at approximately 2,300 metres altitude. Rainfall data is from 1974 to 2004; temperature data is from 1988 to 2004.

beeches (*Nothofagus* sp.). Using these nodal forest types, 'Castanopsis' as lower montane and 'Nothofagus' as mid-montane 'has additional value because it fits more closely with altitudinal zones shown by the birds of New Guinea as well as enabling the major zone of population pressure to be more easily defined for planning of conservation measures' (Johns 1982: 309).

Various highlands groups show as much diversity in describing forest types as ecologists. Among the Nimai, there are two divisions, 'the upper part [is] *dimina*, which they contrast with the lower *same* on the basis of altitude, temperature, crop types and maturation periods, and general vegetation cover. While no sharp boundary separates these two zones, for reference to them is usually relative, the 2,100 m contour line approximates the division' (Hide et al. 1979: 5). Majnep and Bulmer (1977: 35–6) describe five ecological zones that the Kalam delineate: above 2,300 metres is *kamay-agn* 'at the base of the small-leaved southern beech trees', at 2,300–2,100 metres is a 'sweet potato' zone, at 2,100–1,850 metres a 'taro' zone, at 1,850–1,600 metres a 'sugarcane, taro, and banana' zone, and below 1,600 metres is the *nml* zone where *marita* (fruit pandanus, *Pandanus conoideus*) and black palm grow. They note that the zone below 1,600 metres is the one where it is easiest to grow crops. Hyndman (1979) describes a four-part division of environmental zones among the Wopkaimin. As can be seen from this brief survey, groups across the highlands recognise important altitudinal differences in respect to resource management and use.

In Porgera there are three environmental zones that Ipili speakers recognise (Figure 5.1). Below about 1,700 metres is a zone called *wapi*. Similar to the Kalam, this is a zone recognised for its high productivity, but due to dangerous forest spirits it is not used for gardening and living (Jacka 2010). However, it is an area that is used for hunting and foraging. Between about 1,700 to 2,100 metres is the *andakama*, literally 'the place of houses and cleared spaces'. This is the zone in which most gardening is done. Above about 2,100 metres is the *aiyandaka*, literally 'the place of beauty'. It is the high, wet, cold rainforest in which beautification rituals were conducted by young men and women who were preparing themselves for the rigours of life as sexually active, reproductive adults (Biersack 1998, 2004). The altitude boundaries are rather rough estimates as, similar to the Nimai, Porgerans distinguish the zones not on actual altitude, but on a range of factors relating to soil type, temperature, and presence or absence of critical species. Two of the most important markers that delineate the zones are the fruit pandanus, signalling the beginning of the lower zone (as entered from the middle zone), and various cultivars of wild pandanus (*Pandanus brosimos*) that signal the upper zone. GIS analysis that I conducted of aerial photos and satellite data from 1972 supports using 1,700 metres and 2,100 metres as good points at which to distinguish the zones. In 1972 in the eastern Porgera Valley, 78.1 per cent of the total area for gardening was found between 1,700 and 2,100 metres, 14.9 per cent occurred below 1,700 metres, while only 7 per cent of total garden area was found above 2,100 metres. There were no gardens below 1,400 metres or above 2,300 metres in 1972. With the development of mining and the completion of the Enga Highway, this would radically shift by the 2000s, as most occupation occurred above 2,100 metres in the *aiyandaka*. The implications for this transformation are explored later in the chapter.

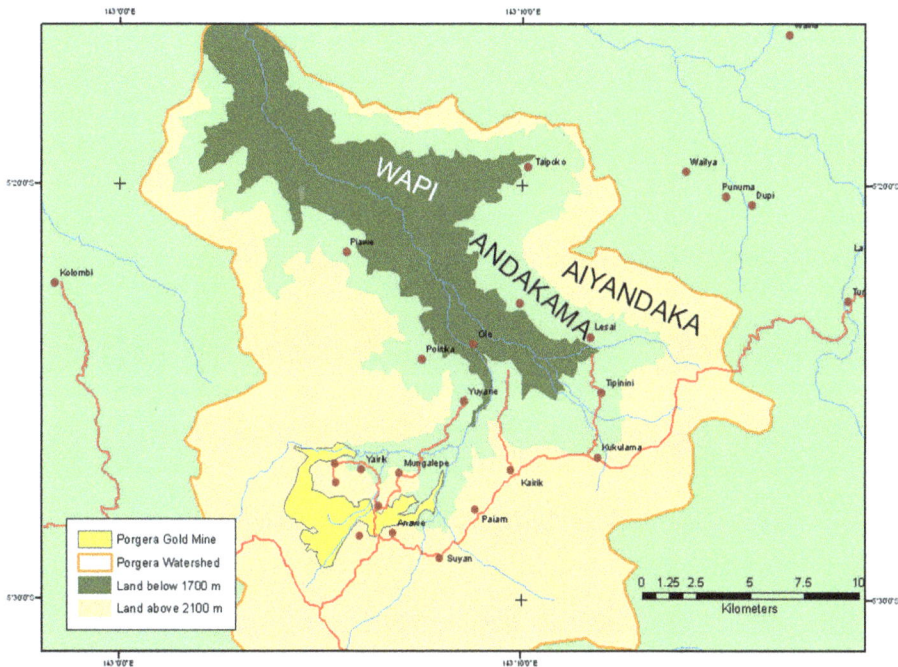

Figure 5.1 Environmental zones recognised by Ipili speakers, Porgera Valley.
Source: Image created by author using ArcGIS.

Social Context and Land Tenure Rules

Porgera is known most famously for its extensive epithermal gold deposits discovered in 1939. Small-scale expatriate mining began in 1948 with local miners becoming active from the late 1950s. The 1960s and 1970s saw a few medium-scale sluicing operations develop alongside continuing local mining, but the planning and development for the large-scale Porgera Gold Mine in the mid to late 1980s fundamentally altered the mining scene in the valley. Since production began in 1990, the Porgera mine has been one of the top gold producers in the world, averaging about 900,000 ounces of gold annually.

Porgera Joint Venture is the operator of the mine. Initially comprising a number of mining companies (most notably Placer Dome, who ran daily operations) and the state, today the joint venture is 95 per cent owned by Barrick Gold and 5 per cent by Mineral Resources Enga, an entity equally owned by the Enga Provincial Government and the Porgera Landowners Association. The Porgera Landowners are members of the seven landowning clans affiliated with the mining lease property.

Since the late 1980s, when mining development began in earnest, outsiders have poured into the valley in hopes of sharing in the wealth generated by the gold extraction (for an overview of issues surrounding the development of the Porgera mine, see Filer 1999; Jackson and Banks 2002). The outsiders are predominantly from neighbouring cultural groups, especially the Huli and Enga, who have long-term kinship and trading links with the indigenous Ipili. Today, I use the term Porgeran to characterise the inhabitants of the valley as so many Enga and Huli have married in, moved in, and built houses and gardens that the social scene today is one of multiple cultures and languages, such that terms such as Ipili, Huli and Enga at best only describe the primary language that some Porgerans speak. However, what is of interest for my purposes in this chapter is to understand how resource rights are determined by a person's link to a landowning group, as rights are not apportioned equally to everyone, especially newcomers.

The basis for social organisation in the valley rests on three principles: land (*yu*), group (*tata* or *yame*) and bilateral kinship principles. In Porgera every named group controls various named blocks of land in the region that are generally demarcated by creeks or ridgelines. What is of importance though is that Porgerans utilise bilateral kin reckoning to determine membership in a group and to regulate property rights. In doing so, Porgeran social organisation ensures that any one person is a member of several groups, generally the eight groups of one's great-grandparents as well as those of their spouse, making 16 groups in all for a married couple to affiliate themselves with. Practically, this would be impossible but it is expected that people will live 'multi-locally' (Biersack 1995), building gardens and houses on two, three or four of their possible kin groups' lands. Most people actually do this for a number of good reasons. A primary reason that people give is that the steep topography, variable soils, and potential for frosts and droughts create a number of different micro-environments such that when one's crops fail in one area, there is a strong likelihood that there will be crops not harmed in another area (Waddell 1975). Another rationale for living multi-locally is the responses that people can make when conflicts break out, such as moving to another group's lands to avoid the fighting and revenge killings that are endemic in this region (Meggitt 1977; Gordon and Meggitt 1985).

There are several terms that recognise the differences in how a person is affiliated to a group, since they do not have equal membership rights in all of their groups, despite the use of bilateral kin reckoning. Male agnates are *tene*. All consanguineous women are called *wana*. The sons and grandsons of women who are consanguineous kin are called *wanaini* and *aiyane lapone*, respectively, and a woman's husband who comes to live on his wife's land is a *wanakalini*. There is a system of incorporation such that multigenerational interaction with a group is recognised through calling a woman's great-grandson *tene*, just as an agnate is called. What this means practically is that if a man goes and lives with his mother's kin and stays there, raising a family, making gardens and helping in compensations, his grandsons will have the same status as male agnates. The rationale for this according to Porgerans is that in terms of the relationships between people and land, in Porgera there were always too few people and too much land. Besides incorporating cognatic kin, Porgeran groups were also keen to incorporate outsiders, called *epo atene*, literally translated as 'those who come and stay'. Ideally *epo atene* would eventually be incorporated as affines through marriage into the group and then within a few generations converted into *tene* (cf. Wohlt 1995). The importance of all these different kin designations relates to how land rights are assigned in the group.

Land among Ipili speakers is conceptualised as having a 'skin' (*umbuaini*) and 'bone' (*kulini*). The skin of the land is the topsoil, plants, trees, structures and gardens, while the bone is all of that underneath the skin. Land is controlled through a communal property system and the bone of the land is said to be 'held' (*mina*) by the group. Individuals within the group 'own' the land's skin (they are called *anduane*, 'landowners') and also inherit land from their fathers and mothers and in turn pass their land to their children. Despite the fact that the Ipili term *anduane* is translated into English as landowner, the true owners are the collective group; individuals merely have long-term and inheritable use rights. Besides inheriting land, one can also gain rights to a piece of land by cutting primary rainforest and then planting a garden or building a house on the land. Even in very old secondary forest, people are aware of who first cut the land (the *anduane*) and anyone intending to make a garden on old secondary forest that is not their own must find the offspring of the actual owner to receive permission to rework the land. This is rather common as land is frequently loaned out to various people for varying lengths of time, which is one way that hosts, the *anduane*, recruit people to come and join the group, hoping that in time they will marry in and their children will become *tenes*. Short-term loans such as these are called *yu no atene* rights, which translates as something like 'land eat stays' or someone who receives land to stay on and plant a garden. All of these forms of affiliation to a clan are important because they determine the kinds of rights that a person has in respect to gaining access to land, converting forest into gardens, and adjudicating conflicts over land. The important thing to note is that these rights are not codified, but are flexible, open to dispute and interpretation, and contingent and context dependent.

Generally, though, *tenes* make most of the decisions as to how land is allocated and make judgements on disputed lands. In terms of a person wanting to cut primary rainforest to build a house or make a garden, they must gain permission from a *tene*. With the exception of *epo atene* and *yu no atene*, any person in any of these categories can do what they like with inherited lands — they can cut trees, plant trees, or even give away portions of land to others. *Epo atene* can give land to outsiders to use; these outsiders (themselves *epo atene*) can't plant trees, although they can cut down trees that are not economically important. All economically important trees in the areas around houses and gardens are individually owned and can only be cut with their owner's permission. Concepts of tree tenure and land tenure are linked in that Porgerans recognise that a lot of effort goes towards maintaining casuarina (*Casuarina oligodon*) and pandanus (*Pandanus julianettii*) trees, and hence anyone who has planted them and maintained them should gain access to the land as well. Therefore, people who loan land must remain vigilant to ensure that trees are not being planted on it, or else risk losing that land. Planting trees is a signal that one intends to claim the land permanently for one's self and one's kin.

Thomas Igila, a Huli man, related a story to me that may well be hearsay, but the message indicates how powerful the concept of planting trees is in terms of land ownership. He told of how a group of Huli in Port Moresby was given a patch of land to garden on by some local landowners. After a while, the Huli asked if they could plant trees on the land and the local landowners allowed them to do so. At that time, the Huli then logically assumed that they were being given the land since they were told they could plant trees on it. When the local landowners asked for the land back after a few years, the Huli refused to relinquish it. According to Thomas, the case went all the way to the Supreme Court of PNG, which found in favour of the Huli. As mentioned, I don't know if a case like this ever made it to the Supreme Court, but as far as local land court cases go in Porgera, people surreptitiously planting trees on land to take it over was a constant source of litigation. In nearly every case, the land is given to the person who has planted trees on it.

In summary, traditionally Porgeran social groups were faced with two facts: abundant land and insufficient numbers of people to defend the group against aggressors. Coupled with unpredictable environmental conditions that challenge their high-altitude subsistence system, and endemic conflict, Porgeran culture responded by having a very flexible social organisation that had numerous mechanisms for incorporating cognatic kin and even outsiders by offering agnatic-like kinship status and land. In the context of mining, though, land has become less abundant. It is still considered important to have a large group, as conflicts are still endemic. One result is that people welcome outsiders, but are less likely to allow them to become true landowners in the sense that they can plant trees and take over lands for their own. Without the right to plant trees, outsiders must then turn to cutting trees in areas of primary forest and secondary forest in order to obtain firewood — enter the firewood crisis of the mid-2000s.

Trees and Land Types

The land types that I use in this chapter — primary forest, secondary forest, gardens and grasslands — are derived from Porgeran terms. Trees are called *ita*, and what I am calling primary forests, Porgerans refer to as *itapu* or *ima*. Gardens are called *e*, and several kinds of gardens are differentiated. *E wene* are 'new' gardens where cultivation has not occurred yet, but trees have been cut down to make the garden; *e siaka*, or an 'alive' garden, is a garden in which sweet potato mounds are being made; *e talu*, or a 'fertile' garden has crops growing in it; *e si* is an old garden with grass growing in it; *e pale* is an old garden with trees growing in it; and *e imati* are forests which show traces of ancient or very old gardening. *E pale* and *e imati* are what I translate as secondary forests. *E pale* are secondary forests that are about five to 25 years old. People generally cannot tell how long ago a garden was in an *e imati* secondary forest, only that it is not *ima*, primary forest that has never had a garden cut in it. Generally people prefer to make new gardens in *e pale*, as soil fertility is greater, but trees aren't as large as they are in *e imati* and *ima*. Grasslands, which I don't discuss here, are another indigenous land type called *puka*.

Within the field of environmental studies there is some debate over how pristine primary rainforests really are (Posey 1985; Parker 1993; Fairhead and Leach 1996; Willis et al. 2004). Balée (1989), for example, argues that approximately 12 per cent of Amazonian *terre firme* forests are anthropogenic. My concern in this chapter is not to argue that primary forests in Porgera have never had gardens in them. The critical point is that Porgerans do not recognise any gardening activity in them that would create large canopy gaps and significantly alter the composition of tree species in *ima* forests. *Ima* forests do serve as supply sources for single logs, for which I have seen plenty of evidence over the years. Therefore, the category 'primary forest' for the purposes of this chapter indicates forests that have received only minor disturbance from humans over several decades.

Culturally, trees figure critically in Porgeran experience. In one origin story, a *tawe akali* ('sky man' spirit) finds a group of women living and gardening in the eastern Porgera Valley. They all are living in a giant water gourd that he picks up and takes to the sky world while the women are sleeping. While they sleep, he lays out food for them and sharpens an *opolanga* (*Schefflera actinophylla*) digging stick for each woman. He wakes the women by playing a bamboo mouth harp (*lino*) and then feeds them and gives them their digging sticks. After eating, he instructs the women to gently dig up a *taiya* (*Araucaria cunninghamii*) tree and plant it next to his house. When they do so, the branches of the tree sprout men whom the *tawe akali* matches up with the women. These couples then become the ancestors of all the clans in the Porgera area, according to the elder woman who told the story. After hearing this story, my research assistant added, 'They became the ancestors of everyone in the world'. In other origin stories, quasi-mythological ancestors plant a number of important tree species as they move about the landscape. Today, these locales still contain dominant numbers of these species marking the passage of these mythical heroes.

Men have lifelong connections to trees. Porgeran etiquette requires that a husband not ask his wife directly the sex of their newborn child, so he discreetly enquires, '*U pe, nu pe?*' which means 'Axe or net bag?' This indicates the cultural implements that define gendered livelihoods in Porgera — axes for men to cut firewood and build houses and fences, net bags (*bilums*) for women to carry produce and babies. Traditionally, after boys were born, the fathers would take the placentas which had been placed in *bilums* and tie them to the tops of very high trees so that as the young men grew up they would not be afraid of heights or of climbing trees to de-limb them. Many men whom I came to know personally or through stories — Mata, Bope, Pawa, Tapaiya, Yakati and others — were, I eventually learned, named after tree species. In time, it often became hard to disassociate the man from the shape or quality of the tree. *Pawa* (*Dacrydium* sp.), a tree known for its strong wood and prickly needles, became entangled with the Pawa who would beat his wife mercilessly for the slightest of infractions. Yakati, the father of one of my elder key informants, Muyu, I never met, but I could never purge my mental image of him as short and broad like the *yakati* 'highland breadfruit' (*Ficus dammaropsis*) tree. Many of the names of places and waterways also come from trees. Porgera comes from the *poketa* tree (*Alphitonia incana*), which presumably grew along the banks of the Porgera River in sufficient numbers to give that river its name. Likewise for the Tupa and Pongema rivers that share names with their associated tree species. There are at least two place names in the eastern Porgera Valley that come from a tree species — both are called Tatonga, meaning 'at the base of the *tato*' tree (*Nothofagus* sp.), a dominant tree in mid-montane forests, as mentioned earlier.

Trees also figure in important rituals both as metaphor and substance. Magic formulas (*kamo*) invoke tree species names to mimic qualities found in those species, such as strength, stickiness, smoothness, watery sap and so forth. To ensure that her husband's skin remained healthy and supple, wives would chew the leaves of *lioko* trees (*Acronychia* sp.) while staring into the sun and reciting a magical spell. Women also buried used moss menstrual pads at the base of *Nothofagus* trees, the largest trees in the Porgeran landscape, which as one male informant indicated were 'the only trees strong enough not to be killed by menstrual blood'. Several rituals designated for healing people and renewing the fertility of the earth required the use of specific tree species to build ritual houses and other structures for their properties of strength and fragrance. Other species are used merely for their decorative purposes; for example, many people employ leaf bundles from species that have green shiny leaves with whitish undersides for the contrastive elements that arise when the leaves are in motion.

Economically, trees of course provide firewood, building materials and fencing. In fact, *ita* means not only tree, but also 'wood' and 'fire'. Perhaps rather than seeing the forest for the trees, trees are seen as 'buildables' and 'burnables'. While some trees are not used for either building or burning, they are used for medicines, rituals and clothing, and as sites for hunting marsupials or birds because of seeds, nuts and fruits preferred by these game animals. *Pandanus* species are also important for food sources and building materials, and the leaves of one cultivar of *P. julianettii* (*tapuna*) are used as compost in sweet potato mounds. In 2006, while gathering data on uses of trees, of 110 species[4] encountered in the field, only one species did not have an economic or cultural use. By far, the most economically important tree is the casuarina (*C. oligodon*), called *yar* in Tok Pisin. Casuarinas are planted around houses and in gardens and grow naturally in disturbed areas, such as near landslips and along eroded stream banks (Robbins and Pullen 1965). Besides being the most important firewood species (Bourke and Harwood 2009: 392) due to its long burn time, high heat output and low ash production, casuarina is also a nitrogen fixer, which contributes to its importance in garden locations. Additionally, it is often used for fencing and house building.

At this juncture, I want to look at data I gathered in 2006 on tree successional dynamics in the different environmental zones in Porgera. While anthropogenically planted casuarina is perhaps the most important tree species, naturally occurring trees in secondary and primary forests are also very important and serve as a critical fallback resource when people need more wood than can be supplied by casuarinas, which is fairly often. To understand

---

4   Overall, Porgerans recognise more than 220 distinct tree species.

what kinds of trees grow where and in what abundances, comparative analyses were done on the composition of tree species found in the three environmental zones and in the differences between primary and secondary forests. As will be seen, altitude has a significant impact on which trees come up in secondary forests, thereby hampering the abilities of Porgerans to obtain adequate wood for building and cooking in the higher-altitude zone.

Methods

Forest Transects

The survey of tree species by land use type and elevation zone was conducted in the eastern part of the Porgera Valley around the Kairik airstrip (5° 26′ S, 143° 10′ E). This was land held by the Bipe clan, among whom I was living in 2006. This site was chosen due to its proximity to the Enga Highway, and because there were several 'camps' of Engans living nearby who were using trees for house building and firewood, allowing me to explore population pressure on local resources and on tree tenure and land tenure dynamics between immigrant and host communities and individuals. Twenty-two 50 × 10 metre (1.10 ha in total) transects were sampled. Transects were established in both primary and secondary forest at each of the three elevation zones, with the exception of primary forest in the upper elevation zone, as there is no longer any primary forest in this area (around the Kairik airstrip). What species should have been in primary forest in the upper elevation zone were reconstructed from oral history and data from other sites in the PNG highlands at a similar altitude.

In the lower elevation zone, three areas (1,500 square metres) were sampled in primary forest (at elevations 1,635–1,652 metres) and four areas (2,000 square metres) in secondary forest (at elevations 1,662–1,684 metres). In the mid-elevation zone, five areas (each 2,500 square metres) were sampled in both primary forest (elevations 2,020–2,091 metres) and secondary forest (elevations 2,070–2,092 metres). In the upper elevation zone, five areas (2,500 square metres) were sampled in the secondary forest only (at elevations 2,357–2,388 metres). The secondary forest sites ranged in age from eight to 12 years since gardens had last been planted in them. Approximate ages of the secondary forest sites were cross-checked with different informants. Initial locations of all 22 plots were chosen by stratified sampling and direction of plot transect was randomly generated, with the exception of ensuring that cliffs and waterways were avoided. All trees with a diameter at breast height (DBH) ≥ 10 centimetres were recorded (Ipili name taken) and measured (DBH).

Focus Groups

Two focus groups were organised to gather data on the cultural and economic uses of all trees recorded from the transect data. One focus group comprising older men met on five occasions, while one focus group of younger men met on three occasions. There was general consensus between the two focus groups on the usefulness of certain species as firewood, building materials and harbouring key animals for hunting, but the elder focus group knew substantially more about the ritual uses of certain trees. The focus groups also discussed what kinds of trees would normally be found in the primary forest of the upper elevation zone. A third focus group composed of both younger and older men was used to list freely the most important trees for firewood, house building and fence construction. Once lists were compiled, individual participants were asked to rank the species in each list in order of importance for the three purposes. Again, ranking exercises showed general consensus within the group of informants as to how valuable each listed species was relative to the others.

Tree Identification and Prior Research

Due to time and budget constraints, voucher specimens were not collected for determination of botanical species in 2006. Species names have been determined by previous work from other researchers and comparison of cognate terms among neighbouring groups. In the 1970s, a dictionary of the Ipili language was compiled by Dr Francis Ingemann and vernacular tree names were recorded and identified in the field by John Womersley, assistant director of the Division of Botany, Department of Forests, PNG (Ingemann n.d.). As well, Ipili and Enga share cognates for several tree species with neighbouring ethnolinguistic groups (Huli and Wola), among whom much species identification work has already been done (Ballard 1995; Sillitoe 1996, respectively). From the entire sample, for all species with five or more individuals, there are only seven species whose scientific names were not determined.

In total, I have conducted 16 months' worth of ethnographic research in the Porgera Valley. Numerous semi-structured interviews on the use of forest resources were held with men and women during a 14-month research period from December 1998 to February 2000, which supplement the findings from the focus groups held in 2006. Forays in the forest with men and women were also a significant focus of research in the earlier research period.

Results from the Ecological Surveys

There is a wide variety in the composition of forests in the highlands; although, as mentioned, ecologists have identified highland forests by 'nodal' species that predominate in certain forests — oaks (*Castanopsis acuminatissima* and *Lithocarpus* sp.) in the lower montane forests and beeches (*Nothofagus* sp.) in the mid-montane forests (Johns 1982; also Paijmans 1976). Porgera fits this characterisation well; oaks comprise 40 per cent of all species in the lower elevation zone, and oaks and beeches account for 40 per cent of all species (11 per cent beeches, 29 per cent oaks) in the mid-elevation zone, which corresponds to the boundary between lower montane and mid-montane rainforest of the ecologists. Across all of the sampled sites, 934 individual trees representing 110 species were recorded. Sixty species (54 per cent of the total) were represented by four or fewer individuals. Twenty-two species (20 per cent of the total) were represented by only one individual. In summary, Porgera's forests have high species diversity but are dominated by a few species. For instance, at the family level, Fagaceae (oaks) is the most prominent family, comprising 29 per cent of all trees. This is followed by Pandanaceae (pandanus) at 11 per cent, Euphorbiaceae (spurges) at 7 per cent and Cyatheaceae (tree ferns) at 6 per cent of all trees.

As would be expected, however, the abundance and distribution of trees in Porgera varies by elevation and forest type (primary versus secondary). Species abundance can be measured in two ways — by the number of individuals of a species per unit of land or by the amount of wood available for use by a species per unit of land. In terms of the amount of wood available for use, foresters use a measure called basal area, which is expressed in square metres per hectare. A final measure that ecologists use to compare the composition of different species in different sized forest plots is importance value (Krebs 1989). Importance value is a relative ranking of species based upon three criteria: (1) species frequency (how many plots a species occurs in divided by total number of plots), (2) species density (number of individuals per hectare) and (3) species dominance (basal area). Values based on these three criteria are relativised (absolute values are expressed as a percentage of the sum of the absolute values) and are ranked on a scale of 0 to 300.[5] As my concern in this chapter is to understand which species of trees are economically important, I focus on importance values to highlight which species in each zone and forest type are frequent, dense and dominant.

---

5   Importance value = relative frequency + relative density + relative dominance. Relative frequency = frequency of one species/total frequency of all species × 100. Relative density = density of one species/total density of all species × 100. Relative dominance = basal area of one species/total basal area of all species × 100.

In the lowest elevation zone, oaks (*Castanopsis* and *Lithocarpus*) are indeed the most important species in both primary and secondary forest. In the mid-elevation zone, *Castanopsis* oaks are the most important species followed by *Nothofagus* beeches. In the upper elevation zone, *Nothofagus* beeches are the dominant species in primary forests based on informants' information, other studies in highlands PNG (Paijmans 1976; Ash 1982; van Valkenburg and Ketner 1994), and from my own work in upper elevation forests in Porgera outside the area sampled in 2006. Of extreme interest is that in upper elevation secondary forests, no individuals of *Nothofagus* ≥ 10 centimetres DBH were found in any of the plots. The most important species were tree ferns (*Cyathea* sp.), followed by a softwooded pioneer species (*Homalanthus nervosus*) and the *tapuna* variety of nut pandanus trees. Table 5.2 summarises the data on importance values in the sampled transects by elevation zone and forest type.

Table 5.2 Trees in each elevation zone and forest type with an importance value > 20.

Elevation zone and forest type	Species	Ipili name	Importance value
Upper secondary	*Cyathea* sp.	*tambu*	30.4
	Homalanthus nervosus	*ambopaiya*	27.4
	Pandanus julianettii	*tapuna*	25.0
Middle primary	*Castanopsis acuminatissima*	*pai*	100.3
	Nothofagus sp.	*tato*	50.1
	Syzygium sp.	*pipi*	26.3
Middle secondary	*Castanopsis acuminatissima*	*pai*	52.4
	Nothofagus sp.	*tato*	25.7
Lower primary	*Castanopsis acuminatissima*	*pai*	68.4
	Araucaria cunninghamii	*taiya*	32.9
	Lithocarpus sp.	*yomboto*	22.2
Lower secondary	*Castanopsis acuminatissima*	*pai*	64.1
	Lithocarpus sp.	*yomboto*	38.9

While the data in Table 5.2 show the most important species in each type of forest and elevation, not all trees are as economically important as others, especially in terms of hardwoods used for firewood, fencing and building materials. Another consideration besides what species are present in each zone and forest type is that the actual amount of wood per forest and elevation is highly variable as well. These results are summarised in Table 5.3.

Table 5.3 Basal area (m²/ha) per forest type and elevation zone.

Elevation zone and forest type	Total basal area	Basal area of hardwood species	Hardwoods as % of total
Upper secondary	17.4	9.6	55.2
Middle primary	78.9	78.6	99.7
Middle secondary	26.3	22.1	84.2
Lower primary	54.8	54.6	99.6
Lower secondary	37.3	36.3	97.3

As seen in Table 5.3, basal area is highest in the primary forests in the mid-elevation zone. This is a function of the size of *Nothofagus* trees found in this zone, many of which had diameters (DBH) of over 100 centimetres. The difference between basal areas in the secondary forest in this zone highlights the absence of exceptionally large trees. In the lowest elevation zone, there are no southern beeches. *Castanopsis* oaks predominate and are much smaller in diameter in general than southern beeches, hence secondary forest does not exhibit such a dramatic difference in change in stand basal area as the mid-elevation zone. The highest elevation zone shows the lowest amounts of useful wood per hectare. No hardwood species were listed as being economically important in terms of firewood, house building or fence construction in the high-elevation zone. However, this zone does provide a greater abundance of resources such as pandanus nuts, pandanus leaves that are used for roofing materials, and tree fern fronds, which are cooked in earth ovens and eaten. Basal area of hardwood species would be higher in high-elevation primary forests than secondary forests (if there had been any left in the study area). In 1999, I observed the clearance of high-elevation primary forest throughout the eastern Porgera Valley. Many of the trees cut were huge, buttressed *Nothofagus* trees. Figure 5.2 shows a *wokabout* sawmill in operation in 1999 in high-elevation primary forest just to the east of the 2006 study site; the stumps and planks are from *Nothofagus* trees.

Figure 5.2 A *wokabout* sawmill in operation in 1999 in high-elevation primary forest, Porgera Valley.

Source: Photo by Jerry Jacka.

Results from the Ethnographic Research

Based on previous research and participant observation, I was aware of how important casuarina was for firewood and fences. For example, in June 1999, I travelled with my research assistant Peter Muyu to the western Enga hamlet of Punuma. Due to various land disturbances that are common to the local area, Punuma sits in a large grove of casuarina trees. Peter and I were to stay with a distant relative of his named Saima. Saima had just built a new house for himself and his new wife and when Peter and I entered the house, Peter made sounds of whistling approval. I assumed it was over the house construction, but Peter elbowed me and pointed to the rafters above the bedrooms that were stocked full of some 50 or more split, dried casuarina logs. Saima just grinned sheepishly at the good fortune he had of living among such extensive casuarina groves.

As mentioned in the Methods section, in 2006 I was interested in trying to get at some sort of consensus over which trees besides casuarina were of obvious economic importance. The categories that my research assistants suggested we should use were firewood, house construction and fence building. Table 5.4 is a composite ranking of the economically important species, as listed by participants.

Table 5.4 Focus group ranking of economically important trees.

Rank	Firewood			Houses			Fences		
	Species	Ipili name	Abundance	Species	Ipili name	Abundance	Species	Ipili name	Abundance
1	Casuarina oligodon	yawale	High	Castanopsis acuminatissima	pai	High	Dodonaea viscosa	lokai	Low
2	Casuarina papuana	yopiyopi	Low	Lithocarpus sp.	lepa	Low	Lithocarpus sp.	yomboto	High
3	Lithocarpus sp.	lepa	Low	Nothofagus sp.	tato	High	Castanopsis acuminatissima	pai	High
4	Castanopsis acuminatissima	pai	High	Compositae	wapeyango	Low	Casuarina oligodon	yawale	High
5	Cunoniaceae	kaina	Low				Unknown	longai	Low
6	Dodonaea viscosa	lokai	Low				Lithocarpus sp.	lepa	Low
7							Unknown	kea	Low

Table 5.5 Number of individual trees per species and importance value forest-wide, based on 1.1 ha plots.

Species	Ipili name	Number of individuals	Importance value
Casuarina oligodon	yawale	n/a	>148.61
Castanopsis acuminatissima	pai	183	148.61
Nothofagus sp.	tato	42	53.02
Lithocarpus sp.	yomboto	46	34.94
Lithocarpus sp.	lepa	12	15.15
Unknown	longai	8	14.08
Cunoniaceae	kaina	6	12.00
Casuarina papuana	yopiyopi	4	10.34
Dodonaea viscosa	lokai	6	9.65
Unknown	kea	1	2.19
Compositae	wapeyango	0	0.00

Of all the species that were ranked as being important, only two appear in all three categories — *pai* (*Castanopsis acuminatissima*) and *lepa* (*Lithocarpus* sp.), both oaks in the *Fagaceae* family. Moreover, only 11 tree species in total were used to rank economic importance in all three categories. Whereas *Castanopsis* is the most abundant tree other than *C. oligodon*, the *lepa* oaks are rather uncommon. For the purposes of abundance, I used the encounter rate across the 1.1 ha plots where 40 or more individuals per species was 'high', 20 to 39 was 'medium', and fewer than 20 was 'low'. Many of the trees that appear in these rankings are low in abundance; it is unknown if this is a factor of being selectively harvested due to their economic importance, or if they are uncommon ecologically. The 11 species from Table 5.4 are ranked using importance values and presented in Table 5.5.

The data in Table 5.5 support the abundance rankings from Table 5.4. The top four species, *yawale*, *pai*, *tato* and *yomboto*, are the only ecologically abundant tree species that are also economically important. *Yawale* (*C. oligodon*) was not used in the calculation of importance values as the numbers of these trees from the 1.1 ha plots do not accurately reflect their actual numbers in the region since people plant them in gardens and around their houses.

Implications

The combined results of the ecological and ethnographic data point to only three species of tree (*pai*, *tato* and *yomboto*), in addition to one species of anthropogenically planted tree, casuarina (*yawale*), as occurring in sufficient enough abundance (in terms of both numbers and basal area) for human exploitation for use as firewood, fencing, or building materials. Only one of these three species, *tato* (*Nothofagus* sp.), occurs in high-elevation forests. However, this species was not found in any of the plots in upper elevation secondary forests, which indicates that overharvesting of parent species or some other factor is limiting its ability to colonise secondary forest settings at high elevations. Mid-elevation forests contain both *pai* (*Castanopsis acuminatissima*) and *tato* (*Nothofagus* sp.) in large numbers. Based on importance values (Table 5.2), primary forests are twice as important as secondary forests for these two species at mid-elevations. *Yomboto* (*Lithocarpus* sp.) is a very marginal tree in mid-elevation forests (importance value of 1.67 on a scale of 0–300). *Nothofagus* is not found in low-elevation forests. Table 5.2 shows that contrary to mid-elevation forests, *pai* and *yomboto* are of more importance in secondary forests at low elevations. For *yomboto*, the difference of importance for secondary versus primary forest is marginal. For *pai*, however, both importance value and basal area are around three times as great for low-elevation secondary forests as compared to primary forests. At this elevation, *pai* seems to flourish with some human disturbance. This is similar to other research on human disturbances of *Castanopsis acuminatissima* around Wau in Morobe Province (van Valkenburg and Ketner 1994). Van Valkenburg and Ketner also looked at the response of *Nothofagus* to disturbance (anthropogenic, landslides, fire, storms, etc.) and concluded that both *Castanopsis* and *Nothofagus* need disturbance to regenerate, but also need mature species to produce seeds. 'At various places ... the forest is no longer able to recover to maturity and if intensity and frequency of human activities increase, species such as *Nothofagus pullei* and *Castanopsis* will also disappear because of lack of seed' (van Valkenburg and Ketner 1994: 53).

These results have great significance given the settlement patterns associated with mining development at Porgera. In the early 1970s, colonial administrators debated the merits of different routes for a road into the Porgera Valley. The main footpath used by colonial patrol officers bisected the most populous portions of the valley for ease of administration and control. A shorter route was eventually chosen that bypassed the main areas of settlement and was located entirely within the high-elevation rainforest at the southern end of the valley (Figure 5.3).

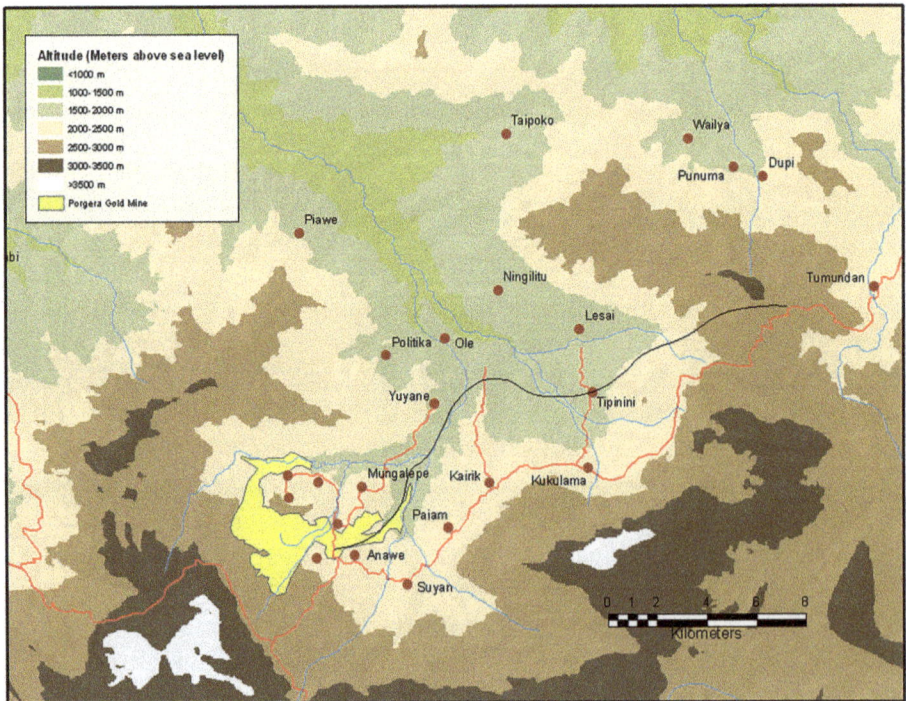

Figure 5.3 The Enga Highway in the Porgera Valley.

Note: The transition in the legend from green to brown delineates 2,000 metres above sea level.

Source: Image created by author using ArcGIS.

Initially, only a few individuals built houses along the road, but with the advent of mining development in the late 1980s, more and more families started to clear forest, build houses and plant gardens adjacent to the road. Around Kairik, big men urged their clansmen to occupy as much land as possible as rumours were circulating that an airstrip to serve the mine would be built next to the road at Kairik. Outsiders were initially welcomed into the area as there was far too much land for the relatively small population. Over the past 20 years, however, population has exploded in the Porgera Valley. From an overall population of 5,011 in 1980, the population increased to 9,255 in 1990 and 22,809 in 2000. Estimated population for 2010 is over 40,000 people. More critically, the population increases have been very uneven, with most of the increases in communities along the road in high-elevation forest. Figure 5.4 shows the rate of population increase in the valley from 1980 to 2000 in the different census units in the Porgera census district.

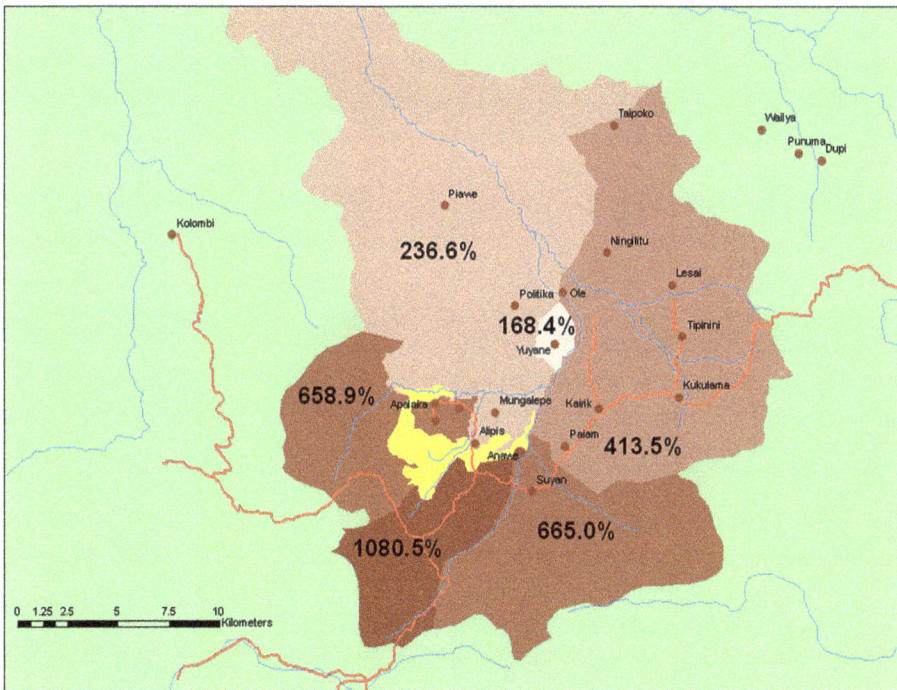

Figure 5.4 Population increase in Porgera census district, 1980–2000.

Note: The area in yellow is the Porgera mine.

Source: Image created by author using ArcGIS.

As seen in Figures 5.3 and 5.4, populations in high-elevation areas of Porgera have grown by approximately seven to ten times their 1980 populations. The area around Kairik where this study was done has experienced a greater than fourfold increase in population. As a result of these patterns, most of the impacts on Porgera's forests have been in high-elevation areas. While these forests can supply the needs of locals and immigrants in their primary states, as demonstrated in the Social Context and Land Tenure Rules section, the secondary high-elevation forests are extremely resource poor with no major hardwood species growing in these forests that can be used for firewood, fences or building materials.

While outsiders were initially welcomed, today land and trees along the Enga Highway are valuable and frequently contested commodities. Numerous intergroup fights and killings have occurred from the late 1990s over the land and resources along the road. Beginning in the mid-2000s, several groups started creating 'forest reserves' in mid-elevation primary forests that could only be used by an exclusive set of clan members who had received permission to cut individual trees in these reserves.

But do the creation of forest reserves, the onset of large-scale mining development and population explosions due to immigration mean that primary rainforest is disappearing in the Porgera Valley? Contrary to the argument by Shearman et al. (2008, 2009) that shifting cultivation is one of the two primary drivers of deforestation of primary rainforest in PNG, McAlpine and Freyne (2001) demonstrate that while PNG's rural population increased by 50 per cent between 1975 and 1996, the total amount of land that went into agricultural production to sustain this population increase expanded by only 11 per cent. As a consequence, rural agriculturists are deploying a range of intensification techniques to increase their yield (Bourke 2001) that are not dependent upon cutting primary rainforest. In fact, Allen et al. (2001: 533) determined the type of land cover that was cleared for planting gardens by rural agriculturalists — 55.8 per cent was secondary forest, 12.9 per cent short woody regrowth, and 30.7 per cent either grass or mixed grass/woody regrowth. Only 0.5 per cent of the land area that was primary forest was cleared for gardens. Of course, Porgera has had vastly different rates of population growth than the nation as a whole. In 1971, a census was conducted which counted 4,020 people in Porgera. In terms of land use in 1972, approximately 448 km^2 were primary rainforest and 25 km^2 were gardens/cleared space. The 2000 census counted 22,809 people, about 5.7 times as many people than in the early 1970s. Land use for gardens more than doubled with 59 km^2 as gardens/cleared space by 2002. The amount of forest between 1972 and 2002 decreased by 36 km^2 to 412 km^2. Population increases in Porgera, therefore, appear to be closely related to loss of primary rainforest; however, intensification is also occurring. For example, in the early 1970s, each square kilometre of garden land supported about 160 people. By the early 2000s, each square kilometre of garden was supporting about 386 people. Of course, the ability to purchase outside food stuffs, mainly rice, certainly influences the latter equation. To summarise for Porgera, both deforestation and intensification are occurring to accommodate the massive population increases into the area. However, as argued in the preceding sections, these population and land use pressures have very localised ecological outcomes as a result.

Conclusions

Porgera's firewood crisis is a result of three interlocking forces resulting from local tenure and resource management practices, colonial history and forest succession dynamics. While Porgerans are active planters of economically important trees, such as *C. oligodon*, and manage forest succession through techniques relying upon shifting cultivation, recent demographic and sociocultural transformations challenge the ability of Porgerans to maintain sufficient forests in certain areas of the valley. Once welcomed outsiders are no longer able to plant trees on borrowed land due to land valuation increasing along the Enga Highway, and also because planting trees on a certain plot of land is a local cultural expression of the intent to claim ownership over that land. Population increases also hamper the effective management of forest succession by limiting the area of land available for shifting cultivation. The sociocultural dynamics of managing trees and forests may not have been hindered if not for the routing of the Enga Highway by outside, colonial decision makers away from mid-elevation hamlets into high-elevation forests. As Porgerans moved higher in altitude to be near the highway, they encountered vastly different conditions for crop production (Jacka 2009) and different forest succession dynamics from mid-elevation forests. While the two most economically important mid-elevation trees besides *C. oligodon*, *pai* (*Castanopsis acuminatissima*) and *tato* (*Nothofagus* sp.), grow well with some anthropogenic disturbances (van Valkenburg and Ketner 1994), secondary forests at high elevations are incredibly resource poor in terms of usable timber for firewood, fences and house building. Trying to balance this confluence of local resource management practices, exogenous historical decisions and ecological factors creates a social, economic, and ecological crisis for most Porgerans. Once-free resources are now commodities and social and ethnic tensions have flared into large-scale resource conflicts (Biersack 2006).

Analysis of the factors underlying the firewood crisis points to the need to approach human–environmental issues as complex, coupled human and natural systems (Gunderson and Holling 2002). The goal of this chapter was not to argue that high-elevation forests in Porgera were 'pristine' and are now under threat from increasing human populations and reduced carrying capacity. Rather, the goal was to examine how human societies and the rainforests in which they live are mutually constitutive (Bayliss-Smith et al. 2003) and how certain internal and external forces challenge the resilience of these human–environmental complexes (Holling 1973). The data I have presented show that human exploitation of mid-elevation forests, while disruptive, is resilient in that key resources in less-exploited forest sections resemble those of exploited sections. The same cannot be said for human exploitation of high-elevation forests in Porgera and probably elsewhere in PNG. This highlights an important policy distinction in that large-scale resource development plans in PNG need to be attentive to the coupled human–natural systems that they intend to impact. Future social and ecological impact assessments need to be more comprehensively synergistic by realising that the societies and environments of PNG comprise one tightly integrated system and not two disparate systems that occasionally interact with one another. Finally, debates about deforestation that focus solely on per annum rates of change or numbers of hectares cleared annually have the potential to overlook the local, culturally contingent and ecologically specific ways that forest resources are used across time in subsistence-based communities.

References

Allen, B., R.M. Bourke and L. Hanson, 2001. 'Dimensions of PNG Village Agriculture.' In R.M. Bourke, M.G. Allen and J.G. Salisbury (eds), *Food Security of Papua New Guinea: Proceedings of the Papua New Guinea Food and Nutrition 2000 Conference*. Canberra: Australian Centre for International Agricultural Research.

Ash, J., 1982. 'The *Nothofagus* Blume (Fagaceae) of New Guinea.' In J.L. Gressitt (ed.), *Biogeography and Ecology of New Guinea*. The Hague: Dr W. Junk Publishers.

Balée, W., 1989. 'The Culture of Amazonian Forests.' *Advances in Economic Botany* 7: 1–21.

Ballard, C., 1995. The Death of a Great Land: Ritual, History and Subsistence Revolution in the Southern Highlands of Papua New Guinea. Canberra: The Australian National University (PhD thesis).

Bayliss-Smith, T., E. Hviding and T.C. Whitmore, 2003. 'Rainforest Composition and Histories of Human Disturbance in Solomon Islands.' *Ambio* 32(5): 346–52.

Biersack, A., 1995. 'Heterosexual Meanings: Society, Economy, and Gender among Ipilis.' In A. Biersack (ed.), *Papuan Borderlands: Huli, Duna, and Ipili Perspectives on the Papua New Guinea Highlands*. Ann Arbor: University of Michigan Press.

Biersack, A., 1998. 'Horticulture and Hierarchy: The Youthful Beautification of the Body in the Paiela and Porgera Valleys.' In G. Herdt and S.C. Leavitt (eds), *Adolescence in Pacific Island Societies*. Pittsburgh: University of Pittsburgh Press.

Biersack, A., 2004. 'The Bachelors and Their Spirit Wife: Interpreting the Omatisia Ritual of Porgera and Paiela.' In P. Bonnemère (ed.), *Women as Unseen Characters: Male Ritual in Papua New Guinea*. Philadelphia: University of Pennsylvania Press.

Biersack, A., 2006. 'Red River, Green War: The Politics of Nature along the Porgera River.' In A. Biersack and J.B. Greenberg (eds), *Reimagining Political Ecology*. Durham (NC): Duke University Press.

Bourke, R.M., 2001. 'Intensification of Agricultural Systems in Papua New Guinea.' *Asia Pacific Viewpoint* 42(2/3): 219–35.

Bourke, R.M. and T. Harwood (eds), 2009. *Food and Agriculture in Papua New Guinea*. Canberra: ANU E Press.

Chapman, C.A. and L.J. Chapman, 2004. 'Unfavorable Successional Pathways and the Conservation Value of Logged Tropical Forest.' *Biodiversity and Conservation* 13(11): 2089–105.

Dambacher, J., D. Brewer, D. Dennis, M. MacIntyre and S. Foale, 2007. 'Qualitative Modelling of Gold Mine Impacts on Lihir Island's Socioeconomic System and Reef-Edge Fish Community.' *Environmental Science & Technology* 41(2): 555–62.

Fairhead, J. and M. Leach, 1996. 'False Forest History, Complicit Social Analysis: Rethinking Some West African Environmental Narratives.' *World Development* 23(6): 1023–35.

FAO (Food and Agriculture Organization of the United Nations), 2005. *Global Forest Resources Assessment 2005*. Rome: FAO.

FAO (Food and Agriculture Organization of the United Nations), 2010. *Global Forest Resources Assessment 2010*. Rome: FAO.

Filer, C., 2010. 'The Impacts of Rural Industry on the Native Forests of Papua New Guinea.' *Pacific Economic Bulletin* 25(3): 135–53.

Filer, C. (ed.), 1999. *Dilemmas of Development: The Social and Economic Impact of the Porgera Gold Mine, 1989–1994*. Canberra: Asia Pacific Press.

Filer, C., R.J. Keenan, B.J. Allen and J.R. McAlpine, 2009. 'Deforestation and Forest Degradation in Papua New Guinea.' *Annals of Forest Science* 66(813).

Gordon, R.J. and M.J. Meggitt, 1985. *Law and Order in the New Guinea Highlands: Encounters with Enga.* Hanover (NH): University Press of New England.

Gunderson, L. and C.S. Holling (eds), 2002. *Panarchy: Understanding Transformations in Human and Natural Systems.* Washington: Island Press.

Hide, R., M. Kimin, A. Kora, G. Kua and K. Kua, 1979. *A Checklist of Some Plants in the Territory of the Sinasina Nimai (Simbu Province, Papua New Guinea), with Notes on Their Uses.* Auckland: University of Auckland, Department of Anthropology.

Holling, C.S., 1973. 'Resilience and Stability of Ecological Systems.' *Annual Review of Ecology and Systematics* 4: 1–23.

Hyndman, D., 1979. Wopkaimin Subsistence: Cultural Ecology in the New Guinea Highland Fringe. University of Queensland (PhD thesis).

Ingemann, F., n.d. 'Ipili Dictionary.' Unpublished manuscript.

Jacka, J.K., 2009. 'Global Averages, Local Extremes: The Subtleties and Complexities of Climate Change in Papua New Guinea.' In S. Crate and M. Nutall (eds), *Anthropology and Climate Change: From Encounters to Actions.* Walnut Grove (CA): Left Coast Press.

Jacka, J.K., 2010. 'The Spirits of Conservation: Ecology, Christianity, and Resource Management in Highlands Papua New Guinea.' *Journal for the Study of Religion, Nature and Culture* 4(1): 24–47.

Jackson, R. and G. Banks, 2002. *In Search of the Serpent's Skin: The Story of the Porgera Gold Project.* Port Moresby: Placer Niugini Ltd.

Johns, R.J., 1982. 'Plant Zonation.' In J.L. Gressitt (ed.), *Biogeography and Ecology of New Guinea.* The Hague: Dr W. Junk Publishers.

Krebs, C., 1989. *Ecological Methodology.* New York: Harper Collins Publishers.

Majnep, I.S. and R. Bulmer, 1977. *Birds of My Kalam Country.* Oxford: Oxford University Press.

McAlpine, J. and D.F. Freyne, 2001. 'Land Use Change and Intensification in Papua New Guinea 1975–1996.' *Asia Pacific Viewpoint* 42(2/3): 209–218.

Meggitt, M.J., 1977. *Blood Is Their Argument: Warfare among the Mae Enga Tribesmen of the New Guinea Highlands.* Palo Alto (CA): Mayfield Publishing Company.

Naughton-Treves, L., D.M. Kammen and C. Chapman, 2007. 'Burning Biodiversity: Woody Biomass Use by Commercial and Subsistence Groups in Western Uganda's Forests.' *Biological Conservation* 134: 232–41.

Paijmans, K., 1976. *New Guinea Vegetation*. Amsterdam: Elsevier Scientific Publishing Company.

Parikka, M., 2004. 'Global Biomass Fuel Resources.' *Biomass and Bioenergy* 27(6): 613–20.

Parker, E., 1993. 'Fact and Fiction in Amazonia: The Case of the Apêtê.' *American Anthropologist* 95(3): 715–23.

Posey, D., 1985. 'Indigenous Management of Tropical Forest Ecosystems: The Case of the Kayapù Indians of the Brazilian Amazon.' *Agroforestry Systems* 3(2): 139–58.

Primack, R. and R. Corlett, 2005. *Tropical Rain Forests: An Ecological and Biogeographical Comparison*. Oxford: Blackwell Publishers.

Robbins, R.G., 1961. 'The Montane Vegetation of New Guinea.' *Tuatara* 8: 121–34.

Robbins, R.G. and R. Pullen, 1965. 'Vegetations of the Wabag–Tari Area.' In *General Report on Lands of the Wabag–Tari Area, Territory of Papua and New Guinea, 1960–61*. Melbourne: Commonwealth Scientific and Industrial Research Organisation (Land Research Series 15).

Scoones, I., 1999. 'New Ecology and the Social Sciences: What Prospects for a Fruitful Engagement?' *Annual Review of Anthropology* 28: 479–507.

Shearman, P.L., J. Ash, B. Mackey, J.E. Bryan and B. Lokes, 2009. 'Forest Conversion and Degradation in Papua New Guinea 1972–2002.' *Biotropica* 41(3): 379–90.

Shearman, P., J. Bryan, J. Ash, P. Hunnam, B. Mackey and B. Lokes, 2008. *The State of the Forests of Papua New Guinea: Mapping the Extent and Condition of Forest Cover and Measuring the Drivers of Forest Change in the Period 1972–2002*. Port Moresby: University of Papua New Guinea.

Shearman, P., J. Bryan, J. Ash, B. Mackey and B. Lokes, 2010. 'Deforestation and Degradation in Papua New Guinea: A Response to Filer and Colleagues, 2009.' *Annals of Forest Science* 67(300): 300.

Sillitoe, P., 1996. *A Place against Time: Land and Environment in the Papua New Guinea Highlands*. Amsterdam: Harwood Academic Publishers.

Struhsaker, T.T., 1997. *Ecology of an African Rain Forest: Logging in Kibale and the Conflict between Conservation and Exploitation*. Gainesville: University Press of Florida.

van Valkenburg, J. and P. Ketner, 1994. 'Vegetation Changes Following Human Disturbance of Mid-Montane Forest in the Wau Area, Papua New Guinea.' *Journal of Tropical Ecology* 10(1): 41–54.

Vayda, A.P. and B. Walters, 1999. 'Against Political Ecology.' *Human Ecology* 27(1): 167–79.

Waddell, E., 1975. 'How the Enga Cope with Frost: Responses to Climatic Perturbations in the Central Highlands of New Guinea.' *Human Ecology* 3(4): 249–73.

Willis, K.J., L. Gillson and T.M. Brncic, 2004. 'How "Virgin" is Virgin Rainforest?' *Science* 304: 402–3.

Wohlt, P.B., 1995. 'System Integrity and Fringe Adaptations.' In A. Biersack (ed.), *Papuan Borderlands: Huli, Duna, and Ipili Perspectives on the Papua New Guinea Highlands*. Ann Arbor: University of Michigan Press.

6. The Structural Violence of Resource Extraction in the Purari Delta

JOSHUA A. BELL

The Papuan forests teem with valuable timbers, choice gums, medicinal herbs, aromatic spices, and a hundred products which, to win, need only the educated efforts of the aboriginals, for the enrichment of the world's markets (Bevan 1890: 284).

Introduction

On the bank of the upper reaches of the Aivei River sat an empty cargo container, detritus from a failed logging and oil palm venture initiated in 1993 (Filer with Sekhran 1998: 188–9; Bell 2009). In 2002, when I first encountered this container, the forest was slowly engulfing its blistering orange-red surface (Figure 6.1). All the heavy equipment brought to this site had been removed or scavenged for parts, and the empty cargo container was all that remained. In 2010, when I returned, the container had slipped from the bank and was now submerged in the river. While every forest has its share of the remnants of failed projects, the forests, mudflats and tidal creeks of the Purari Delta, and the wider Gulf of Papua, have more than their fair share. Their presence and lingering absence serve as a testament to the failed and ongoing intersection this region and its inhabitants have with capitalism, and the extent to which forests are central to these engagements.[1]

---

1   See Mawyer, Chapter 2, this volume, as well as the evocative explorations by Raffles (2002), Kosek (2006) and Gordillo (2011).

Figure 6.1 Empty container enveloped by the forest on the banks of the Aivei distributary of the Purari River.

Source: Photo by Joshua A. Bell, 2002.

This container possesses companions to the west on the bank of the Pie River at Baimuru government station. One has sat next to the police station for at least 20 years, and until 2010 was used as the holding cell (Figure 6.2). This is where, following verbal rebukes and typically a beating by the police, the accused were kept for a night to several days before being freed. More often than not, the occupants of this cell were young men charged with the cultivation and possession of *kuku dipi* (marijuana), the making of 'steam' (home-made alcohol) and having an illegal home-made gun. Since 'zoom' (outboard motor fuel) is too scarce or costly to be used for transporting prisoners to the provincial capital (Kerema) or the national capital (Port Moresby), these young men would either return to their village or remain hanging out in Baimuru, becoming a lingering source of anxiety for all. Collectively, these young men, and the lawlessness they are perceived to embody, are understood to be emblematic of the new era of consumption, jealousy and the ascendancy of individualism over kin relations, known locally as *moni kaeou* ('money ground') (Bell 2006a).[2]

---

2 Indicative of this new order, the first murder occurred at Baimuru station in December 2005 around Christmas. The victim, a man in his late 40s, is said to have been killed while walking home in the early morning by young men he angered at an all-night card game (Kaia Rove, personal communication, 6 April 2006). Gambling has become a new norm in the delta as communities become temporarily awash with money through royalty payments (Bell 2009).

Figure 6.2 Container turned into a jail, now disused.
Source: Photo by Joshua A. Bell, 2010.

Further south on the river's bank lie another set of containers that have been converted into the base for a watchtower from which locally hired security guards sit waiting for both real and imagined threats (Figure 6.3). The watchtower helps guard the operations of ethnic Chinese merchants affiliated with the Malaysian multinational Rimbunan Hijau, which has conducted the bulk of the logging operations in the Papuan Gulf. These merchants came to the delta on the invitation of the current parliamentary member for the Kikori Open seat, who owns the plot of land on which the watchtower has been built. Their operations are part of the larger eclipsing of locally owned and operated businesses. The success of this operation has made it and the ethnic Chinese operators the objects of much jealousy. The intensity of this jealously emerged most starkly in 2005 when 'rascals' (*raskols*) came by outboard motor from Kerema, robbed the store, and almost killed its manager, Jackson. When I met him in 2006, Jackson gladly showed me the scar that radiates down from his shoulder along his back where a bush knife split him open. Following the elaboration of this encounter in Tok Pisin, he quickly changed the conversation to ask whether I had any United States currency to exchange with him for Papua New Guinea (PNG) kina.

Figure 6.3 Containers on the bank of the Pie River used to form a security tower outside a Chinese-run trade store in Baimuru.
Source: Photo by Joshua A. Bell, 2010.

Condensed in this detritus of failed and ongoing resource extraction, and the ongoing wakes that these encounters help generate, we have vestiges of Tsing's (2005) 'frictions', where the assemblage of local, global, human, non-human, nature and culture are materialised and reconfigured (Raffles 2002). These and other containers in the delta are 'good to think' not only because of the different scale-making projects they help enact and foreground (Tsing 2005: 57), but also due to the metaphorical and relational aspects of 'cargo' (*kago*) by which Melanesians remake the terms of their engagement with the economic and moral inequities of the global economy of resource extraction (Wagner 1981). If the first container evokes failure and the lurking quality of things submerged, then the second and third are evocative of the misunderstandings, violence and transactions, legal and illegal, that occur as these projects unfold. All conjure up the entropic effects of the delta and the ongoing attempt by outsiders to wrest resources — whether they be coal, oil, timber, sago, copra, rubber or eaglewood — from its enduring flow. These activities have directly and indirectly involved inhabitants of the delta in the reworking of place and subjectivity, and are part of the cutting and reknitting of the networks by which the delta as both real and imagined place is connected and disconnected.

Building on my earlier work on the informal economies around resource extraction and the transformative effects of incorporated land groups (Bell 2006b, 2009), I focus here on the conjoined effects and ruptures of resource extraction and the various types of violence they engender. Specifically, I draw on Farmer's articulation of 'structural violence' (Farmer 2004), through which he attempts to understand the conditions and effects of inequality in and on communities. A key aspect of this violence is the erasure of history and understanding of the relations that engender this oppression.[3] Formulated in response to the particular conditions of Haiti, structural violence can be seen to be part and parcel of what Harvey (2005), reformulating Marx, refers to as 'accumulation by dispossession'. This inequality — whether economic, social or representational — is an increasingly salient aspect of neoliberalism and capitalism more generally, and has profound effects along the supply chain, which in turn feed back and impact local communities. Structural violence is particularly visible in the frontiers of capitalism where resources are made and unmade, and relations and histories erased (Tsing 2005; Kosek 2006). As we shall see, structural violence and its effects are a key force in the shaping of people's understandings of the assemblage of relations that are critical to the formation of their environmentality (Agrawal 2005: 166). I am specifically interested in how the various forms of violence have shaped I'ai people's responses to outside efforts to tap their forests for the world's markets, and the corresponding transformations of their knowledge about the forest, others and the self that these engagements have involved. In this way, I seek to contribute to discussions of how the state and companies articulate their relations with communities through violence (Dinnen and Ley 2000; Wood 2006; Lattas 2011), with the full acknowledgement of how difficult writing about such violence is, and the dangers in both overstating and understating its slippery dimensions (Scheper-Hughes and Bourgois 2004).

Moving through a brief history of resource extraction in the Purari Delta, I proceed to a discussion of the wakes of recent events, memories and current processes. In what follows, I do not make an attempt to provide a comprehensive history of Purari communities' engagement with resource extraction; rather, I offer some salient events that show how relationships have been transformed over time.

---

3   Farmer's formulation builds explicitly on the work of the Latin American theologian Johan Galtung (1969), and has clear parallels to the articulations of 'everyday violence' (Scheper-Hughes and Bourgois 2004) and 'symbolic violence' (Bourdieu 1991), which are equally concerned with the power dynamics of lived reality and the misrecognition of the causes of inequality.

Histories of Resource Extraction in the Purari Delta

While my focus is primarily on events since 2000, it is important to situate these contemporary processes within longer-term and fractured histories of resource extraction in the region. These intersecting histories are products of what Theodore Bevan (1890) termed 'the enrichment of markets'. Credited with being the first European to discover and travel up the Purari River, Bevan's remarks in this chapter's epigraph are prescient of the later ventures that came to the region. His optimism resonates with these projects and foregrounds the ways in which the Purari Delta and its surroundings have been reconfigured as resources through a variety of ventures: the Vailala oil fields (1912), Wame sawmill (1922), surveys by the Australasian Petroleum and Anglo-Persian companies (1920s–1930s; 1950s–1960s), the proposed Wabo dam (1970s), Rimbunan Hijau's logging ventures (since 1995), InterOil's prospecting and drilling (since 2002), and resurgent interest in the Purari hydropower project (2010). Moreover, Bevan's relentless collecting of flora, fauna and ethnographic specimens is emblematic of the bundling of activities carried out by most Europeans in the region, whereby various entities became resources to be extracted and sold or collected for science (Welsch et al. 2006; Bell 2013).

While far from uniform, these resource extraction projects over the past 100 years have worked to transform communities' relationships to their forests as the colonial and post-colonial state has sought to gain access to the region and extract timber. To extend a metaphor drawn from the reality of the delta's environment, this process continually involves ebbs and flows, and periodic submerging and revelation of what has been covered. As a result, this process is uneven at best, and is re-articulated anew with each generation and each new project regionally.

Following Bevan's initial incursion into the Purari Delta, the region slowly came under European influence. In 1905, Reverend John Henry Holmes of the London Missionary Society inhabited Urika station on the delta's coast (LMS 1906: 318). Alongside his evangelising, Holmes also sought to transform communities through work. He began employing Purari boys to manufacture rattan chairs, which they sold within Papua and in Australia. Until the 1940s, *kaia ini*, as the rattan was known, was also used in the construction of the central ritual figure, the *kaiaimunu*. The construction of these wicker figures was central to the empowerment of male initiates (Williams 1923a; Bell 2010). While it is unclear if Holmes understood the symbolic associations of *kaia ini*, these industrial activities were central to the belief of Holmes and other London

Missionary Society members that spiritual salvation could be achieved through hard work. The symbolic analogies at play here most likely did not escape Purari people's attention.

Alongside these uses of forest materials, Holmes also embarked on modification of Urika's immediate landscape. Holmes quickly encountered the nesting of relationships that Purari communities have to place. As Holmes remarks, upon settling in Urika, '[we] unwittingly … had stepped into a totemic hornet nest; we were stung on every hand, but were innocent of the provocation' (Holmes 1924: 145). He comments:

> We had to fell trees, cut canals and demolish on one hand that we might build up on the other. Our proceedings alarmed people in all the villages; they were concerned because the island, being an *awapa-mako* [*opa mako*, place of spirit-beings], their totem-kinship demanded of them approval or disapproval, of our doings. (Holmes 1924: 145–6)[4]

Holmes's comments are suggestive of the conflicts that arose around cosmologically important places for the Purari people during this early period, as well as their perceptions of subsequent transformations enacted by later projects.

Far from being an undifferentiated tidal swamp, the delta was and still is understood to be a palimpsest of culturally significant places whose meanings are constituted through both past and present practices and contemporary narrative. *Airu omoro*, a class of narratives, speak of these connections understood as *kapea* (paths) by which past actions of *imunu* (ancestral spirit-beings) are inscribed in the landscape connecting people to their wider world (Bell 2006b). Prior to the iconoclasm of the Tom Kabu Movement (1946–69), communities materialised their relations to *imunu* through large-scale rituals and displays of objects, which acted as the *ruru* ('skins') of these otherwise immaterial beings (Williams 1924; Maher 1961; Bell 2010). Accompanied by feasting, these rituals helped affirm traditional leadership positions and relations within descent groups. The *ravi* (longhouses) that housed this material were quite literally forests of relations transposing, through carved images and masks, ancestral presence with various sensorial and social effects. With the abandonment of the *ravi* and with few heirloom objects left (Bell 2009), the remaining *imunu* are understood to reside in specific trees, whirlpools and stones and become visible through taking on different human, animal or mechanical forms.[5] Consumption habits

---

4   In 1913 Holmes gives a sense of the scope of these endeavours: 'Side by side with building our station, canal-making, land-clearing for gardens to support a family ranging from 150 to 300 during the 5 yrs a coconut plantation of 6000 young palms was planted and in another year the first fruits from this plantation will be gathered' (CWM/LMS 1913).

5   There are three types of *imunu*: *kae* ('earth') *imunu*, *ere* ('water') *imunu* and *iri* ('tree') *imunu*. I was told that *iri imunu* are the most like humans.

and subsistence activities, as elsewhere in Melanesia, further root people into this system whereby the growth of one's children's bodies is understood to be a manifestation of ancestral efficacy and parental nurture. Christianity (Seventh-day Adventist, Pentecostal and United Church denominations) has helped to further attenuate communities' engagement with, and understanding of, these beings. For many, the forest has become a place of danger where demonic forces now dwell (Brunois 1999). These beings pioneered the *kapea* that form the present landscape that underlie social formations and their histories, all of which have become central to the Purari people's negotiations about resource ownership (Bell 2009).

Within a year of the establishment of Urika, the administration created a government station at Kerema from which it could launch semi-regular patrols into the delta (Murray 1912: 177). When the area was deemed pacified after 1908, the administration sponsored several expeditions to follow on initial discoveries of coal and oil in the Papuan Gulf (Carne 1913; Wade 1914; see also Rickwood 1992: 38–42). These surveys resulted in the identification of the Vailala oilfields in the neighbouring Gulf Division in 1912. This site, along with Papua's nascent plantation economy, needed new sources of labour, so recruiters turned to the delta.[6] Labour increasingly involved men in unprecedented travel out of the delta to plantations in the neighbouring Gulf and Central districts, as well as to work in Port Moresby. During these sojourns, men were exposed to Papuans of various cultural backgrounds, the Australian work regime, and the evolving colonial language of Police Motu. While the London Missionary Society mission condemned labour recruitment because of its effects on their own missionisation efforts (LMS 1915: 328), the colonial administration felt that labour was beneficial (Murray 1912). This domestication of Purari communities through labour, laws and missionisation was part of the broader attempt to domesticate communities and harness their resources.

Purari communities were not passive participants within the emerging colonial economy, but engaged in a series of tacit and explicit negotiations about the nature and extent of their participation. The so-called 'Vailala Madness' of 1919 dramatically brought the simmering regional disenchantment with the colonial economy, and the inequality of the moral relationships it engendered, into sharper focus. The explosion of economic activity and the new valuations of regional resources helped contribute to this event (Williams 1923b; Bell 2006b). While temporarily affecting communities on the coast, the effects of the Vailala

---

6   The 1907 Native Labour Ordinance allowed individuals to be indentured for up to three years. All contracts were signed in the presence of the district's resident magistrate or assistant resident magistrate. The resident magistrate received the labourers' wages monthly and then witnessed their being paid in full upon the end of the contract. In the event of a labourer's death, he would pay out the wages to relatives in the village (Lewis 1996: 49).

Madness soon passed, as new colonial policies, namely the Native Taxes Ordinance (1919) and the Native Plantations Ordinance (1920), prompted communities to participate more fully in the commercial transactions.[7]

During this decade several new regional commercial ventures gave Purari communities more direct access to trade goods and became local sources of employment (Territory of Papua 1923: 53; Hope 1979). The most successful and long-lasting commercial venture established was the Wame sawmill and trade store. Established in 1922 by Lewis Lett, a successful trader and planter in Gulf Division, Wame quickly became an economic and social hub for the district (Woodward 1922: 1). The sawmill processed timber until 1958, after which it was moved to the newly established station of Baimuru in 1961.[8] By 1925 the sawmill employed 65 villagers from all over Delta District and was home to three Europeans (Liston-Blyth 1925: 2). By 1943 the sawmill employed 350 labourers (Ross 1943: 4) and was sending timber to Daru and Port Moresby (Chance 1937: 2). While trees were already sites of valuation that were 'generative or productive relations between persons' (Leach 2004: 42), the sawmill's activities helped trees become a new locus by which to create and sustain relationships with Europeans, and indeed within Purari communities themselves. For example, on 22 August 1926, the assistant resident magistrate, C.H. Karius, records that 75 landowners from the Baroi villages of Akiaravi, Oravi and Evara were paid £90 (or just over £1 each) in return for the right to cut timber along the banks of the Baroi River over a period of 10 years (Karius 1937: 2–3). These funds entered into the local economy in the form of trade goods. Such commercial opportunities augmented those carried out around sago, and allowed for the furthering of asymmetric power relations between elders, who held hereditary chieftaincies, and the young men who struggled to obtain trade goods through their labour. While colonial threats of violence often forced men into these labour situations, as much as their wanderlust, on their return they faced sorcery reprisals within

---

7   Taxes were imposed on all males between 18 and 36 and placed in a special fund used only for projects that were deemed beneficial to the Papuan, such as the funding of schools, hospitals and technical training (Lewis 1996: 149). The Native Plantations Ordinance set aside land near villages for plantations, and the government provided tools and seeds, while the villages supplied their labour. The resulting profits were used to pay their annual tax, and helped instil villagers with a sense of industry. In the early 1920s coconut plantations were established near every major village in the delta, and villagers were encouraged to try planting a range of produce (Territory of Papua 1923: 50). While it is difficult to trace the impacts of these policies, they not only induced men to more readily sell their labour, but most likely also induced them to sell their resources of timber, sago and fish.

8   In 1937 Steamships Trading Company purchased a half interest in the operation. During World War II the mill was taken over by the Australian military administration for timber production for the war effort, although production slowed due to the lack of boats to transport milled timber to Port Moresby (Bennett 2009: 103). After the war, the operator was Port Romilly Timber Mills (Nicklason 1969: 243–48).

their community if they proved unwilling to distribute their goods.[9] This form of social disruption, one of the tacit results of resource projects escalating the uneven distribution of wealth, is ongoing in the delta, and remains one of the most destabilising effects of these projects (Bell 2006a).

This structural violence was accentuated by the venereal diseases that these capitalist incursions helped to spread. Although present in 1912, the increased rate of infection in 1922 was such that the administration established a medical outpost near the mouth of the Aivei River (Territory of Papua 1923: 116). The increased transmission in the delta was deemed a result of the traffic of women from the Urama and Goaribari communities for trade goods and shell valuables (Woodward 1921: 1), but no mention is made of European involvement. As with the present circumstances of the Papuan Gulf, logging and oil exploration camps created new commodity contexts in which prostitution flourished as communities without direct access to European goods attempted to access them. As recalled by I'ai elders in 2001 and 2002, Goaribari groups continued to come to Purari villages until 1947, and to the camps into the 1960s. The structural inequalities that helped give rise to the trade of sexual services for trade goods has continued into the present. Since 2000, it has not been uncommon for communities seeking to access the flow of real and imagined wealth around present-day resource extraction projects to seek to marry their daughters to men working in the camps and to obtain materials from the logging ships (Bell 2006a; Hammar 2010).

Following World War II, the region was transformed by the Tom Kabu Movement, which combined iconoclastic fervour with intense interest in economic independence through cash cropping (Maher 1961, 1984). Although suppressed by the Australian administration, this movement helped to open up communities to the possibilities of new things, and shift the cultural expectations and horizons of communities. Soon afterwards, the Australasian Petroleum Company (APC) began conducting surveys in the delta's hinterland and to the west in Kikori (Hicks 1953: 695–700; Pym 1958). In the 1950s, the APC established a base at Middletown, at the juncture of the Sirebi and Kikori rivers, and by 1955 had four rigs operating. It spent £3 million a year and had a labour force of 'thousands' of men through Delta District and beyond, with 20 vessels, numerous launches, helicopters and so forth (Hope 1979: 249). Following these unsuccessful surveys, in 1960, the Consolidated Zinc Corporation and the British Aluminium Company began investigating the

---

9   During a patrol in 1914, the resident magistrate of Delta Division reported that, in the village of Pai-ara, 'young boys that go to work, on their return are obliged to hand over some of their hard earned trade, to appease the old scoundrels who profess to be Sorcerers ... their dread of the Sorcerer is so great, that some of the boys as soon as they arrive, go to the Sorcerer, and give him some of their trade without being asked for it by the Sorcerer, this is done to ensure that the Sorcerer will leave them alone' (Ryan 1914: 5–6).

upper Purari to initiate a hydroelectric scheme. In 1972 the Japanese company Nippon Koei conducted another preliminary study of the feasibility of a dam, which in 1973 was reviewed by the Australian company Snowy Mountains Engineering Corporation. Then in 1974 an American consortium — led by Bechtel Engineering, General Electric and Westinghouse — expressed interest in this project as a source of power for uranium enrichment, ushering in a new wave of planning studies centred on a proposed 2160 megawatt dam. While environmentalists in Australia and PNG protested against this scheme (Pardy et al. 1978), it also failed the test of economic feasibility, and was abandoned.

Following a lull in the copra market, in the 1990s, the Papuan Gulf came into focus again as multinational conglomerates began reinvestigating the region's potential for logging and oil production with new technologies. A conjunction of factors in the 1990s — specifically, the increased demand for hardwoods and PNG's fiscal crisis — encouraged both international interest in harvesting these forests and government eagerness to raise revenue by granting concessions (Filer 1997; Filer with Sekhran 1998; Filer et al. 2000). The Purari River constitutes the boundary between two major logging concessions. To the east is the Vailala Block 3 concession (200,100 hectares), which has been controlled by Rimbunan Hijau since 1992 and operational since 1999. To the west is the Baimuru Block 3 concession (438,300 hectares), which was allocated to Turama Forest Industries in 1995, but where logging operations had not yet commenced at the time of writing.[10]

In 2001–02, Rimbunan Hijau subsidiary Frontier Holdings carried out operations in Vailala Block 3 from its original base camp at Kaumeia on the Purari River. The site of this camp is held by the I'ai to be the ancestral site where the female ancestors Keia and Auei paused on their way to find the moon. Over its five years of operation (1999–2004), the camp stripped this site bare, turning it into an eroding patch of earth. The second camp on the Aivei branch of the Purari River near Evara village was opened in 2004. A third camp, simply known as the Purari camp, was opened in 2007 and was still operating in 2012.

This scenario shifted dramatically with the activities of the Canadian company InterOil that began exploratory drilling in the Purari Delta's hinterland in 2002. Despite InterOil's drilling activities remaining spatially remote in 2006, its presence on the upper Purari River continued to loom large in the imaginations of coastal villagers. This was fuelled in part by glimpses of large helicopters carrying equipment inland and by movement of the company's barges up and

---

10   Permits to Turama Forest Industries (TFI) were first issued in 1988, and in 1995 extended to cover 1.7 million hectares for a duration of 35 years. There were rumours linking TFI to Rimbunan Hijau for many years, but these were only confirmed in 2011, when the Rimbunan Hijau subsidiary Niugini International Corporation Ltd was officially recognised as operator of the Turama concession.

down the river. As of 2006 InterOil possessed three petroleum prospecting licences (PPL 236, 237 and 238), which covered coastal and inland areas in Central and Gulf provinces. The upper Purari River falls within PPL 238. Since 2009, InterOil's activities have been the source of incredible speculation in the delta communities. In March 2010 I was told of how the company's new base camp being built above the former logging camp at Kaumeia would rival Port Moresby and lift the delta out of obscurity and on to the centre stage of national development. While lobbying for InterOil's proposed overland pipeline to follow the Purari River, community members also talked of an impending meeting to reinitiate the Wabo hydroelectric project on the upper Purari River. In September 2010, the Australian company Origin Energy Ltd and PNG Sustainable Development Program Ltd (PNGSDP) announced a memorandum of understanding to commence the Purari Hydro Scheme (Barrett and Elks 2010). This project was abandoned when the PNG government took over the PNGSDP shares in the Ok Tedi mine in 2012.

The Ebbs and Flows of Violence[11]

The recent wave of resource extraction and exploration has exacerbated existing tensions in communities and created new problems due to perceptions of unequal royalty payments (Bell 2009). As these projects have unfolded, the state has become increasingly absent. The exception, however, has been the bursts of periodic presence of the mobile squad, who without exception have come to enforce the movement and activities of the logging company. As elsewhere in PNG (Dinnen and Ley 2000; Dinnen 2001; Wood 2006; Lattas 2011), this has distorted people's views of the state and its interests in protecting communities, and helped enforce a growing sense of communities' alienation from their land. To this end, a major campaign promise of the winning candidate in the 2002 Kikori Open election was how, if he was elected, Rimbunan Hijau would fill the void of the state and provide communities with health clinic buildings and roads. Here I recount some aspects of these encounters as they are configured in, and around, a police officer known as *Inamu Monovae* ('One-Eye').

Violence around the current wave of logging appears to have begun in September 2000 when inhabitants of Aumu village near the mouth of the Aivei River held up a small cargo ship headed to the then-active logging camp of Kaumeia. Due to complications prior to and following the 1997 elections, the I'ai had been at

---

11   Although presented in a coherent chronological narrative, the events I recount only became clear to me in 2010 with the benefit of hindsight and interviews with community members of Aumu. In October 2000, when I first came to the delta, I had heard about these troubles but until 2010 had not had the chance to interview participants.

this time frozen out of the project's royalty payments (Bell 2009). Already frustrated by the lack of compensation for the use of their waterways, the men also stopped the vessel because of increased erosion caused by the ships' wakes and the pollution these boats are thought to cause by dumping their bilge. Men took eight pallets of food as payment and then let the ship proceed. A mobile squad from Moresby was dispatched once the incident had been reported and Aumu was raided. I was told how the chief, councillor and teacher were asked to line up and then told to jump into the river and swim to the far bank and back while the police shot the water nearby. Villagers' canoes were riddled with bullets, and all the dogs, ducks and chickens were killed. The terrified villagers ran away and lived in their fishing camps in the forest for a month. They were brought back by the son of a policeman, whom they credit for saving the village from further reprisals. However, for some time, subsequent cargo ships would go by the village at full speed and fire a warning shot or two as they passed. At night these ships would also rake their high-powered spotlights over the village to wake and scare people. Recalling these events in 2010, one man remarked, 'Aumu is the only one fighting for this river!'

This incident was one reason why villagers throughout the Purari Delta were terrified when the mobile squad returned on two separate occasions in 2001 and 2002 (Bell 2006a: 220–1). My return to Baimuru in October 2001 coincided with the arrival of a mobile squad detachment that had not only been dispatched to deal with disturbances at the Kaumeia logging site, but was also raiding villages in a search for illegal home-made firearms and marijuana. The rumours circulating that the police were assaulting women on the coast were such that men in Mapaio talked of how they would meet the police with force if necessary. At the centre of these rumours was One-Eye, a man of Oro and Milne Bay parentage who served during the war in Bougainville. His nickname derives from the patch he wears over an eye damaged in this conflict. Within the delta it is widely believed that One-Eye possesses X-ray vision in this covered eye, which makes him all the more fearsome. More than this, however, narratives also circulated in 2001 and 2002 that, during his persecution of youth in Baimuru, One-Eye berated them in Tok Pisin by saying that his actions were payback because 'your grandfathers ate my grandmothers, now you will eat dirt!'[12] When asked about the incident, one elder related how his uncles told him of an event that occurred in Oro Province near Buna during World War II, when they were serving in the Papuan Infantry Battalion and had helped Australian commanders gather and kill Oro men who had collaborated with the Japanese. They were then said to have raped the men's wives and sisters.[13]

---

12   *A, tumbuna bilong yu kaikai tumbuna bilong me, nau yu kaikai graun!*
13   I have never been able to ascertain if there is any truth in this remembrance. It may be that the story recalls the Higaturu trials of 1943 during which Papuan collaborators were hanged.

This man thought that One-Eye was using the police action as a way to enact his own revenge for this historical event and, as his story quickly spread, this only added to local people's fear of the police commander.

In September 2002, One-Eye returned to the delta with a detachment of six men to investigate and retrieve seven chainsaws that had been stolen from the Kaumeia camp. Upon hearing of their imminent arrival in Mapaio, where the I'ai were preparing to host their Independence Day celebrations, the local councillors held an impromptu meeting. Some 30 men gathered in the dark on the village's rugby field to talk about what might happen. The talk that followed revealed their fear of One-Eye and the police more generally, as well as the structural inequalities that logging had brought about in people's access to cash and store-bought goods. The meeting began with one man speaking in the local language:[14]

> One-Eye is going to do damage. One-Eye told the workers at Kaumeia base who are holding the chain that he wants to go inside and check if the chain is there or disappeared. While One-Eye went inside to see how far the chain came out from the bush, he wanted to come out and ask the workers questions again. But the workers were frightened and they ran away. They had never seen police. But he did not come to beat them, he just came to check their rooms. The policeman was just asking them, if some say yes they will get pain, but they ran away. When they come to the village it will be a big problem. So these policemen coming will spoil the village. Policemen are saying the village has no trouble in it, but these boys are coming so there will be trouble. The names of the people are in their hands, but if you two councillors are not here they are going to really spoil us …

> One-Eye came down and beat one man at base. That man got all his things and brought out everything in [a] clear place. That is what happened at base. What people did he brought it out. That man was from somewhere else [another area of PNG]. Our small ones were just staying there, and that fellow they belted said everything. Our small ones got frightened and came down. They should have stayed there but they got frightened and came down [to the village].

The speaker and others present urged the councillors to settle the matter before the police came in force and caused the young men to flee into the forest, raising suspicions and most likely leading to their being beaten. One young man was suspected of having fled the camp with a chainsaw and, on hearing of the arrival of the mobile squad, had already fled Mapaio. Another man at the meeting interjected:

> You people have not seen the action of One-Eye. I have seen his actions when I went with … to Epea [Vailala base camp]. He is not going to care about anybody, he will do whatever he wants. One-Eye is not going to care about anyone. He will

---

14   The following texts have been translated into English with the help of another local man.

chase the councillors too because the police sent a message for … who stole [the] company chainsaw for it to be put in the councillor's hands and return it, but you people didn't do it. They have sent [a] message already to do that. But this policeman doesn't care. When he wants to beat people, he will beat them. He will beat the councillors too. This is what I saw in Epea [Vailala].

To this a third speaker added:

The police are coming, we don't know what their thinking is. Some of us don't know how they beat people, some of us go around Moresby and know … but they are still coming and they are going to be here in the village. So we must think and do the things for the health of the people. Police is coming down, and our children, boys, women and children are going to get pain by running into the bush. They will come and not look to the councillor but hit him and the chief.

As the meeting went on talk whirled about how the villagers had brought this problem on themselves by unduly pressuring their sons and nephews working at Kaumeia to provide them with store-bought food or asking to stay with them while marketing produce at the camp. People also spoke of the evils of *kuku dipi* and how marijuana was to blame for the stealing of the chainsaws. Invoking more fears of the possibilities of what might happen, a fourth speaker said:

Please if you are father of the house, don't run away because these people are dogs and will do whatever they want. These people are dogs. You don't leave the house if they come in the night, because if you leave the house your wife and daughter will be smashed by them. So just stay in the house, don't leave.

Adding fuel to these fearful speculations, a fifth speaker, a former policeman, remarked on how he had once raided a village in the highlands:

A chopper came up and left us in the middle of the village. When we went to the house there was no one. We burned the house, and just shot the dogs, cats and pigs. They will come down and do this … they did not sleep good and so they are angry and they are going to come down and do these things. So you young ones must think and do the right things for old ones.

As the night wore on, tensions between men who did and did not have kin working at the logging camp worked themselves out. The councillors urged calm and promised to resolve the matter by finding the stolen chainsaw reportedly located in the village, and that any confrontations would not result in violence. At 2:00 am the group dispersed, still fearful that the night or dawn might bring a raid.

The next day the mobile squad arrived at 9:30 am with One-Eye leading the group. Their arrival caused all sports being played in celebration of Independence Day to stop. Only one young man had fled on the news of their

arrival. The councillors met with the squad and were given a list of suspects —
24 young men who all worked at Kaumeia. While the councillors sat with the
Independence Day committee to discuss the names, One-Eye asked me over to
talk. Speaking in English, he remarked how he had heard I was here but wasn't
sure what I was doing. After I explained my work, One-Eye commented that he
was on a routine patrol of the area during which he and his men had confiscated
some marijuana up in the border Pawaiian area. He also explained that this
operation was about recovering chainsaws, which cost K2,000 each. Two had
been recovered but five were still missing. One-Eye then remarked: 'It's for
logging — um — because those machines are the ones that are used to make
money for landowners. Without those machines the company can't cut logs …
We are coming through the leaders in an easy way.' Then, referring to himself,
he noted that 'Papuans are peace loving. Oro Province are like highlanders, they
need force.'[15]

Following our brief conversation, One-Eye addressed the villagers assembled on
the rugby field in a mixture of Tok Pisin and English:

> There are independence celebrations in Gulf, Western Province, Milne Bay,
> Southern Region too. Oro is where they are strong headed, so when I go [there]
> I use force like in the highlands. Now we don't come like that this morning. I don't
> want the community to be scared, so I came to your leaders in a good way. I want
> you to have a normal Independence Day celebration because plenty of villages
> have come to celebrate. I want all these people to enjoy PNG independence.
> So that is why I come in a good way. Now all these people they don't know why
> the police are come. They have just come to celebrate independence. You have all
> come to cooperate, which I appreciate. Now you need to work with your leaders.
> Whatever we do please cooperate with what we need so we can leave you to your
> celebrations. Because we could have come and raided your village at night but
> I don't want to do this because you need to celebrate your Independence Day.
> That is why we came true through to your leaders. Through the leaders we can
> talk and work with them to find what we need. Force is a last resort, huh? Some
> places this isn't the case but you people are peaceful. I just want to tell you all
> this so you can cooperate.

Appealing to purported regional variations in temperament, and playing up
what could have happened, One-Eye tried to position himself as a peaceful
enforcer of the law. My unconfirmed suspicion was that he had not raided the
village in part due to my presence. The chainsaw believed to be in Mapaio was
eventually recovered and the young man responsible taken to Baimuru where, in
addition to being beaten, he was held in the cargo-holding cell for several days
before being released. In the end, the visit was a peaceful one but, as evident

---

15 My notes are from memory following our conversation. Oro Province is technically part of the Papuan
region, but One-Eye seems to have thought of it rather differently.

in the above commentary, it provoked an outpouring of anxieties about the violence that could come — violence that the men had seen and heard about being inflicted elsewhere. As discussed elsewhere in PNG, these rumours not only create new structures of fear, but they also distort a community's sense of the state, and can significantly reshape what people believe is possible and permissible (Jacka 2001; Kirsch 2002; Butt 2005; Wood 2006). These distortions are constant features of the shifting configurations that are the politics of place and value that resource extraction helps enact (Tsing 2005).

I next encountered One-Eye at the third Rimbunan Hijau logging camp on the Purari River in March 2010. In 11 years since the first camp was established at Kaumeia, logging roads (passable only during the dry season) had connected this site to a logging depot at the mouth of the Vailala River. Moreover, with the economic opportunities present in the logging camp, I had heard how women often visited camps bringing items for sale to workers and their families. Indeed, when visiting the camp, I encountered three Southern Highlands women in the market who were on their third trip down to sell store-bought goods. It was only on my second trip to the camp that I met One-Eye. He saw me and emerged from underneath a house where he was sitting, smoking and speaking with others. After learning what I was up to, and recalling that we had met before, he asked to speak to me out of earshot, and we proceeded to talk for the better part of an hour. He explained how he and some companions had come three weeks ago to establish a police base and how they were conducting periodic patrols. Telling me of his recent patrols to stop marijuana trafficking from Kerema, he also alluded to his membership of a special task force to help ensure law and order in the region, particularly in light of the InterOil project, which he assured me was of special interest to the United States. Earlier in my trip, I had been to the new base that InterOil was building and had met two very bored mobile squad policemen. Armed with automatic weapons like One-Eye himself, it was their job to stop and investigate all river traffic. However, as my I'ai friends explained, their lack of a searchlight meant that it was very easy to slip by the checkpoint under the cover of darkness.

While One-Eye's stature seemed rather diminished by the time we met in 2010, he was still feared by the I'ai and other Purari communities. His reputation for brutality continued, despite the more limited range of his actions in patrolling the logging roads. But if he and his companions at the InterOil base represented the PNG state, and by extension the companies they ultimately served, their presence was not without friction. As suggested in the next two examples I discuss, narratives abound about how the environment itself is a lively actor pushing back against the actions of these individuals and companies (Tsing 2005: 29).

Figures of Resistance

In the late afternoon in March 2010, I sat on the veranda of John's house drinking tea.[16] My companions and I had come to the settlement of Kae Varia after surveying the upper arm of the Ivo branch of the Purari River from Mapaio. During this trip we had visited the former logging camp at Kaumeia, now green with secondary growth, and had also stopped at the Pawaiian settlement of Poroi before coming down the eastern arm of the Purari. At Poroi we encountered a once-vibrant village reduced to a single household. All the inhabitants had moved up to Wabo to be closer to the activities and possibilities around InterOil's operations there. This trip up the river was an attempt to fulfil a promise I made to the Mapaio community in 2006 to follow up on the survey of the Purari River I had first completed in 2002. Their and my hope was that I would produce a map for the community with ancestral sites and landmarks marked out through a global positioning system. Then and now, communities were deeply concerned about the destruction of ancestral sites by Rimbunan Hijau and now InterOil, and collectively there was a belief that a map could help make their claims legible (Peluso 1995). Indeed, already we had seen evidence of this destruction obliquely through being shown stone tools that workmen had either unearthed themselves or found on the roads as they were being cut.

In 2003, concerned about the direction in which things were headed, John and some 50 men and women had moved to this site on the west bank of the east branch of the Purari River from the Aikavalavi section of Mapaio. I had worked with John's father during my initial fieldwork, and this visit provided us with an important opportunity to catch up after eight years. At the time of my visit his father was convalescing in the regional hospital of Kapuna and died shortly thereafter in 2011. Moving to this spot on the strength of his father's ancestral connections, John and his companions were part of a contingent of I'ai who have moved up the Purari River to have more immediate access to the informal economies of the project areas, as well as to assert their ownership of land that was otherwise not normally inhabited.[17]

---

16   My return to the Purari Delta in March 2010, after a four-year absence, was shaped by the unexpected death of my adopted father and principle research assistant Kaia Rove. His death was understood to have been connected to ongoing leadership disputes within his clan. 'John' is a pseudonym.

17   At Ne'ea swamp, a traditional fishing and hunting locality, we came upon another group of I'ai from Mapaio who had established a semi-permanent camp on the riverbank in 2006. They had done so to be closer to the Kaumeia logging camp, and to have access to the InterOil project. Prior to their arrival, however, they had been visited by InterOil officials and police telling them that they had a month to vacate the project area or they would be forced to leave. Further down the river is the I'ai village of Vikoiki, an offshoot settlement from Aumu, another I'ai village located further south on the Purari River. Established in 2001, Vikoiki is on the west bank of the Purari River, opposite a now abandoned logging loading point used by Rimbunan Hijau. Indicative of the desires and misunderstandings that the logging and for that matter InterOil's activities are generating, a group of people from Muro (a village site north of Orokolo Bay) has occupied the abandoned logging point since 2008. Their presence and claims to the site have generated a lot of resentment from the residents of Vikoiki who claim the area as theirs.

The conversation on John's veranda turned to the conditions at the Purari logging camp ten minutes downriver. I had visited the Purari logging camp earlier in the day, spending six hours meeting with workers to ask them about their work conditions.[18] They were understandably guarded in their conversations with me, and flatly denied that they had any issues with gambling, prostitution, alcohol or marijuana, as documented elsewhere in PNG (Hammar 2010). When I raised this with John and company, the group started laughing and began enumerating the activities that took place in the surrounding area.[19] John commented how *kuku dipi* 'is the popular one! [Laughing] ... if you were there at night you would find them on the road'. The group said that gambling flourished and that prostitution also occurred. Our talk turned to how the camp had been damaged by fire in March 2009, and the structures I saw had only recently been rebuilt. The only undamaged buildings were the Seventh-day Adventist church and the Asian workers' quarters, which are at some distance from those of the local workers.

By one account of this event, a number of men and some women were gambling, drinking and listening to music in the single men's barracks on the night of the fire. As the revellers carried on, out of the forest and into the building walked a naked white woman. Stunned, the group is said to have watched her walk through and then out of the building. Shortly thereafter the fire broke out and spread wildly. The other story was that someone's small stove had got out of control while boiling water during a late-night gambling session. Men routinely have small kerosene stoves for cooking in their sleeping quarters, where they also hide zoom. Some workers sell stolen fuel on the black market to supplement their marginal incomes, which are all too quickly spent buying overpriced foodstuffs and other consumables in the camp's store. It is thought that the fire accidently ignited a store of this fuel.

While the second story speaks to the grinding conditions of the logging camp, the first story is an example of a particular genre of narrative that I had been hearing since my arrival in the Purari Delta in 2000. These stories focus on how *imunu* periodically surface and disrupt the activities of the living. The cumulative effects of resource extraction regionally have only further exacerbated the breakdown of people's relations with *imunu*. Indeed, local people tell of how *imunu* have either been taken by foreigners, as in the case of one spirit-being that took the form of a stone, or have been driven out by activities like seismic

---

18   This trip was a follow up on some work I had conducted in the Kaumeia logging camp in 2001 and 2002. On this visit, I interviewed three workers — a grader of roads and two chainsaw operators from Morobe Province.

19   Informal economies flourish both within and alongside these camps due to the captive market of the workers and their isolation. While visiting the next day, we ran into a relative of one of my companions from Iuku village in Orokolo on the coast. The relative had come to the camp to sell betel nut to workers at a huge margin. He remarked how he also sold rum when it was available.

testing. Large trees are the main home of *iri imunu*, which makes logging the most disruptive of these activities. In the past, and for those men still conscious of tradition today, protocol dictates that the principal male present addresses the *imunu* that resides within a large tree before it is cut down to make a house post or canoe. The *imunu* will be told of the group's intention to cut down their home, and then asked if they will leave. Food is offered to the *imunu*, and is subsequently consumed by the men. Once the meal is over, the cutting of the tree can commence.

Awareness of these relations within the Baimuru community of Kararua was such that a large feast was held before logging began in their western territories in 1999. During the feast the *imunu* were asked to find new homes and not to harm the community. The I'ai, with whom I principally worked, never did anything similar, which has made them persistently wary about working at the logging operations and sensitive to the problems that non-placated *imunu* can cause. As a result, my I'ai friends saw the white woman as the forest manifesting itself and wreaking vengeance on the morally dubious workers. In 2001 and 2002, rumours circulated of how the tutelary ancestral figure Ivia and the still-living but hidden figure of Tom Kabu were variously behind the endless deluge of rain. The narratives that abounded about these actions directly linked them to local people's exclusion from royalties or compensation payments (Bell 2009).

It was during this trip that I also heard of a *kae imunu* named Aupea who, in the shape of a stone, blocked InterOil's barges carrying equipment to Wabo. According to my acquaintances, the *imunu* had, like the white woman, decided that enough was enough, and was actively blocking the barges. His actions were so successful that the barges had to navigate a different river passage to avoid his movements. In these narratives, these *imunu* become spectral weapons of the weak, disrupting frontier capitalism and opening up spaces through which the I'ai can hope for recognition.

A Scarf Wrapped around the Head

While the structural violence of resource extraction finds unique expression in the examples I have discussed thus far, it is also found in more quotidian examples. In closing I want to turn to two examples drawn from experiences in Kararua and Mapaio. Separated by eight years, these two examples point to the enduring inequalities engendered by the ongoing resource extraction in the region.

Just south-west of Baimuru government station, Kararua village is occupied by a confederation of clans that compose the Vaimuru tribe. Regionally, the Vaimuru were the strongest adherents of tradition, but had been receiving timber royalty payments since 2000. This periodic influx of money had increased gambling in the village and was a growing source of domestic strife. This strife took a variety of forms, from kin yelling at one another during the distribution of payments, such as I observed one afternoon, to the inability of the community to come together with contributions of sheet iron to cover one communal building for the sake of an increased water supply. Instead, indicative of their fragmentation, each household had their solitary sheet and catchment. The most profound embodiment of these increased disruptions was in a scarf that an older woman, Kiki, wore tied tightly around her head. I had come to know her husband, Avae, and when visiting Kararua I always made an effort to spend time with them both. On one of my visits, Avae and I began speaking about intergenerational problems caused by logging. Our talk quickly turned to gambling, and then to Kiki's scarf. Avae and Kiki disclosed how her scarf kept a skull fracture in place, and how Kiki was racked with headaches without it. One day, as they gambled with newly dispersed royalty payments, one of their sons ran out of funds. On being refused a loan by his mother, he hit her with a wooden plank. The family had since mended the social strain that this violence had caused, but Kiki continued to suffer the legacy of this encounter. Avae's anger had not abated, and he had all but concluded that he would die keeping all his knowledge of his descent group's lands to himself.

In 2010, when I returned to Mapaio, I found that gambling also dominated life in the village, yet people were even more outspoken about the pollution of the Purari River. This is the main source of drinking water for the community, and a major source of food (fish, crayfish, etc.) and water used in the processing of food (sago). For years community members had noticed that sedimentation in the water had increased, the taste had deteriorated, and there were more ailments associated with unpalatable water, such as intestinal complaints and more frequent and lasting sores. Indeed, eye infections and body sores, which were previously confined to the wet season, were now a constant occurrence. One small girl's sore was so bad that her toe had to be removed. Alongside these health issues, fish were increasingly found covered in sores.

One possibility is that these sores may be due to the effects of increased population in the central highland provinces. In his discussion of aquatic pollution discerned from testing in 1978, Petr (1983: 329) notes that the Purari catchment receives wastewater from the major towns of Goroka, Kundiawa, Minj, Mt Hagen and Mendi. Increased faecal material from the greater animal and human populations in the catchment could be increasing these various health and environmental problems. However, Petr notes that the Purari River at Wabo appears to be self-purifying and, in any case, pollution of the main river should not be such a great concern because 'people along the Purari prefer drinking rainwater and water from side streams' (ibid.: 331). While this preference still exists, the reality is that many communities do not have an alternative to drinking the water from the main river.

It is also likely that the higher sedimentation load is caused by deforestation both in the delta and in the highlands. While it remains to be seen whether InterOil's activities are having the sort of effects on the Purari River that are experienced by the community of Mapaio, the company's community affairs officers have warned communities against drinking the river water or eating dead fish found in the river. They did not link these restrictions to the company's own activities, but one cannot help but speculate on the relationship.

Conclusion

> If place-making is a way of constructing the past, a venerable means of *doing* human history, it is also a way of constructing social traditions and, in the process, personal and social identities. We *are*, in a sense, the place-worlds we imagine. (Basso 1996: 7)

In this chapter, I have attempted to think through the place-worlds that resource extraction in the Purari Delta helps to imagine, and with it think through the structural violence these activities give rise to and help enact. Working in this region, one is very aware of the legacy of earlier representations, how writers seeking to bracket and frame the Purari people portrayed them as primordial cannibalistic semi-aquatic mud dwellers (Bevan 1890: 186–8; Murray 1912: 197; Holmes 1924: 28–9). While the violence of these representations is only obliquely felt locally (Peluso 2003), spectres of this past do periodically appear. I have tried to expose aspects of the micro-politics of structural violence in the Purari Delta, show how resource extraction structures life at the cost of local communities, how companies use police forces to make communities comply with their wishes, and how communities themselves become more violent in the wake of the new instabilities of wealth and power that resource extraction brings (Lattas 2011). In doing so, I hope not to paint too bleak a picture of the Purari Delta where, despite the problems I outline, people find the resilience to joke, laugh, sing and enjoy aspects of their lives. However, the situation in the delta, as in other regions affected by such projects, is stark and not to be understated (Kirsch 2006; see other contributors to this volume). My hope is that in bringing together these histories, and contemporary accounts, more attention will be brought to bear on the impacts of resource extraction regionally, and that the affected communities may begin to obtain the justice they deserve and the compensation they need.

When I asked a long-term friend and Mapaio resident, Henry, about the way he saw the future for his children in light of these events in 2012, he remarked:

> Like I have said, myself I am not going to stop [it] because the people they want the development to go in. The government wants the development to go in. If I a village person goes to stop the development then the government is going to come against me and beat me again. So that is why the village they want to stop the particular project for the future but the government is forcing that thing to happen. So we have no power to stop it. We have concern for our children there to stop it but we have no power because the government has power to do everything there. So our government is not concerned about histories they are just going in and doing these things inside.

References

Agrawal, A., 2005. *Environmentality: Technologies of Government and the Making of Subjects*. Durham (NC): Duke University Press.

Barrett, R. and S. Elks, 2010. 'Papua New Guinea River to Power Up North.' *The Australian*, 16 September.

Basso, K. 1996. *Wisdom Sits in Places: Landscape and Language among the Western Apache*. Albuquerque: University of New Mexico Press.

Bell, J.A., 2006a. 'Marijuana, Guns, Crocodiles and Radios: Economies of Desire in the Purari Delta.' *Oceania* 76(3): 220–34.

Bell, J.A., 2006b. 'Losing the Forest but Not the Stories in the Trees: Contemporary Understandings of the Government Anthropologist F.E. Williams' 1922 Photographs of the Purari Delta, Papua New Guinea.' *Journal of Pacific History* 76(3): 191–206.

Bell, J.A., 2009. 'Documenting Discontent: Struggles for Recognition in the Purari Delta of Papua New Guinea.' *The Australian Journal of Anthropology* 20(1): 28–47.

Bell, J.A., 2010. 'Out of the Mouths of Crocodiles: Eliciting Histories with Photographs and String Figures.' *History and Anthropology* 21(4): 351–73.

Bell, J.A., 2013. ' "Expressions of Kindly Feeling": The London Missionary Society Collections from the Papuan Gulf.' In L. Bolton, N. Thomas, L. Bonshek and J. Adams (eds), *Melanesia: Art and Encounter*. London: British Museum Press.

Bennett, J.A., 2009. *Native and Exotics: World War II and Environment in the Southern Pacific*. Honolulu: University of Hawai'i Press.

Bevan, T.F., 1890. *Toil, Travel, and Discovery in British New Guinea*. London: K. Paul, Trench, Trubner.

Bourdieu, P., 1991. *Language & Symbolic Power* (ed. J.B. Thompson, transl. G. Raymond and M. Adamson). Cambridge (MA): Harvard University Press.

Brunois, F., 1999. ' "In Paradise, the Forest is Open and Covered in Flowers." ' In C. Kocher Schmid (ed.), *Expecting the Day of Wrath: Versions of the Millennium in Papua New Guinea*. Boroko: National Research Institute (Monograph 36).

Butt, L., 2005. ' "Lipstick Girls" and "Fallen Women": AIDS and Conspiratorial Thinking in Papua, Indonesia.' *Cultural Anthropology* 20(3): 412–42.

Carne, J.E., 1913. 'Notes on the Occurrence of Coal, Petroleum, and Copper in Papua.' Melbourne: Department of External Affairs (Bulletin of the Territory of Papua 1).

Chance, S.H., 1937. 'Kikori Patrol Report No. 8 of 1936/37, 10–22 March 1937.'

CWM/LMS (Church of World Mission/London Missionary Society), 1913. 'Papua Letters, Box 14, Folder 1, Jacket B, 18 April 1913 (GB0102).' London: School of Oriental and African Studies Library.

Dinnen, S., 2001. *Law and Order in a Weak State: Crime and Politics in Papua New Guinea*. Honolulu: University of Hawai'i Press.

Dinnen, S. and A. Ley (eds), 2000. *Reflections on Violence in Melanesia*. Canberra: Asia Pacific Press.

Farmer, P., 2004. 'An Anthropology of Structural Violence.' *Current Anthropology* 45(3): 305–25.

Filer, C. (ed.), 1997. *The Political Economy of Forest Management in Papua New Guinea*. Boroko: National Research Institute. London: International Institute for Environment and Development.

Filer, C. with N. Sekhran, 1998. *Loggers, Donors and Resource Owners*. London: International Institute for Environment and Development in association with the PNG National Research Institute.

Filer, C. with N.K. Dubash and K. Kalit, 2000. *The Thin Green Line: World Bank Leverage and Forest Policy Reform in Papua New Guinea*. Canberra: The Australian National University, Research School of Pacific and Asian Studies.

Galtung, J., 1969. 'Violence, Peace, and Peace Research.' *Journal of Peace Research* 6(3): 167–91.

Gordillo, G., 2011. 'Ships Stranded in the Forest: Debris of Progress on a Phantom River.' *Current Anthropology* 52(2): 141–67.

Hammar, L., 2010. *Sin, Sex and Stigma: Pacific Response to HIV and AIDS*. Wantage (UK): Sean Kingston.

Harvey, D., 2005. *The New Imperialism*. Oxford: Oxford University Press.

Hicks, E.G., 1953. 'The Search for Petroleum in the Territories of Papua and New Guinea.' *South Pacific* 7: 688–700.

Holmes, J.H.R., 1924. *In Primitive New Guinea*. London: Seeley, Service & Co.

Hope, P., 1979. *Long Ago Is Far Away: Accounts of the Early Exploration and Settlement of the Papuan Gulf Area*. Canberra: Australian National University Press.

Jacka, J., 2001. 'Coca-Cola and *Kolo*: Land, Ancestors and Development.' *Anthropology Today* 17(4): 3–8.

Karius, C.H., 1937. 'Kikori Patrol Report No. 7 of 1937/38, 8–11 December 1937.'

Kirsch, S., 2002. 'Rumour and Other Narratives of Political Violence in West Papua.' *Critique of Anthropology* 22(1): 53–79.

Kirsch, S., 2006. *Reverse Anthropology: Indigenous Analysis of Social and Environmental Relations in New Guinea*. Stanford (CA): Stanford University Press.

Kosek, J., 2006. *Understories: The Political Life of Forests in Northern New Mexico*. Durham (NC): Duke University Press.

Lattas, A., 2011. 'Logging, Violence and Pleasure: Neoliberalism, Civil Society and Corporate Governance in West New Britain.' *Oceania* 81(1): 88–107.

Leach, J., 2004. 'Land, Trees and History: Disputes Involving Boundaries and Identities in the Context of Development.' In L. Kalinoe and J. Leach (eds), *Rationales of Ownership: Transactions and Claims to Ownership in Contemporary Papua New Guinea*. Wantage (UK): Sean Kingston.

Lewis, D.C., 1996. *The Plantation Dream: Developing British New Guinea and Papua, 1884–1942*. Canberra: Journal of Pacific History.

Liston-Blyth, A., 1925. 'Kikori Patrol Report No. 1 of 1925/26, 2–5 August 1925.'

LMS (London Missionary Society), 1906. *Reports of the London Missionary Society, No. 111: From April 1st, 1905 to March 31st, 1906*. London: Alexander & Shepheard.

LMS (London Missionary Society), 1915. *Reports of the London Missionary Society, No. 120: From April 1st, 1914 to March 31st, 1915*. London: Alexander & Shepheard.

Maher, R.F., 1961. *New Men of Papua: A Study in Culture Change*. Madison: University of Wisconsin Press.

Maher, R.F., 1984. 'The Purari River Delta Societies, Papua New Guinea, after the Tommy Kabu Movement.' *Ethnology* 23(3): 217–27.

Murray, J.H.P., 1912. *Papua or British New Guinea*. London: T. Fisher Unwin.

Nicklason, N., 1969. 'The History of Steamships Trading Company Limited.' In K.S. Inglis (ed.), *The History of Melanesia*. Port Moresby: University of Papua New Guinea. Canberra: The Australian National University, Research School of Pacific Studies.

Pardy, R. et al., 1978. *Purari Overpowering PNG?* Fitzroy (Melbourne): International Development Action for Purari Action Group.

Peluso, N.L., 1995. 'Whose Woods Are These? Counter-Mapping Forest Territories in Kalimantan, Indonesia.' *Antipode* 27(4): 383–406.

Peluso, N.L., 2003. 'Weapons of the Wild: Strategic Uses of Violence and Wildness in the Rain Forests of Indonesian Borneo.' In C. Slater (ed.), *In Search of the Rain Forest*. Durham (NC): Duke University Press.

Petr, T., 1983. 'Aquatic Pollution in the Purari Basin.' In T. Petr (ed.), *The Purari: Tropical Environment of a High Rainfall River Basin*. The Hague: Dr W. Junk.

Pym, L.A., 1958. 'Oil Exploration in Papua: Transport Problems.' *Walkabout* 24(2): 10–14.

Raffles, H., 2002. *In Amazonia: A Natural History*. Princeton (NJ): Princeton University Press.

Rickwood, F., 1992. *The Kutubu Discovery: Papua New Guinea, Its People, the Country and the Exploration and Discovery of Oil*. Glenroy (Australia): Frank Rickwood.

Ross, J.C., 1943. 'Kikori Patrol Report No. 2 of 1943/44, 9–20 August 1943.'

Ryan, H.J., 1914. 'Kikori Patrol Report No. 16 of 1914/15, 24 November – 8 December 1914.'

Scheper-Hughes, N. and P. Bourgois, 2004. 'Introduction: Making Sense of Violence.' In N. Scheper-Hughes and P. Bourgois (eds), *Violence in War and Peace: An Anthology*. London: Blackwell.

Territory of Papua, 1923. *Annual Report for the Year 1921–22*. Government of the Commonwealth of Australia.

Tsing, A.L., 2005. *Friction: An Ethnography of Global Connection*. Princeton (NJ): Princeton University Press.

Wade, A., 1914. *Petroleum in Papua*. Melbourne: Government Printer for State of Victoria.

Wagner, R., 1981. *The Invention of Culture*. Chicago: University of Chicago Press.

Welsch, R.L., V.L. Webb and S. Haraha, 2006. *Coaxing the Spirits to Dance: Art and Society in the Papuan Gulf of New Guinea*. Hanover (NH): Hood Museum of Art and Dartmouth College.

Williams, F.E., 1923a. 'The Pairama Ceremony in the Purari Delta, Papua.' *Journal of the Royal Anthropological Institute of Great Britain and Ireland* 53: 361–87.

Williams, F.E., 1923b. *The Vailala Madness and the Destruction of Native Ceremonies in Gulf Division*. Port Moresby: Government Printer.

Williams, F.E., 1924. *The Natives of the Purari Delta*. Port Moresby: Government Printer.

Wood, M., 2006. 'Kamula Accounts of Rambo and the State of PNG.' *Oceania* 76(1): 61–82.

Woodward, R.A., 1921. 'Kikori Patrol Report No. 2 of 1921/22, 21 July – 4 August 1921.'

Woodward, R.A., 1922. 'Kikori Patrol Report No. 7 of 1922/23, 23–31 October 1922.'

7. The Fate of Crater Mountain: Forest Conservation in the Eastern Highlands of Papua New Guinea

PAIGE WEST AND ENOCK KALE

Introduction

Papua New Guinea's (PNG's) Crater Mountain Wildlife Management Area (CMWMA), a conservation-as-development project that began informally in the late 1970s – early 1980s and that was solidified by national and international conservation policies and practices in the 1990s, effectively ceased to exist in its original form in March 2005. The CMWMA, the oldest Wildlife Management Area (WMA) in the country, was a 2,700 square kilometre area located at the borders of Eastern Highlands, Simbu (Chimbu) and Gulf provinces. The area that the CMWMA encompassed is home to the Gimi and Pawaia peoples who believe that their day-to-day lives and social relations with the living and dead bring their forests and the plants and animals in them into being. The lands that the CMWMA encompassed matter to Gimi and Pawaia because they sustain them and are sustained by them, because they hold and tell their history, and because they are the source and the sink for their cosmological relations with the past and the future (see Ellis 2002; West 2005b, 2006a). Although neither of these sociolinguistic groups think of the world in terms of western notions of 'value', conservation scientists, activists and practitioners saw Crater Mountain and the landscape around it as biologically valuable — 'a natural resource of national and global importance' (Johnson 1997: 394). They saw it as valuable for three main reasons. First, much of it is covered with forest that is highly biologically diverse with high rates of endemism. Second, the area encompassed by the WMA was large enough to cover the landscape between lowland rainforest on the Purari River and montane cloud forest on Crater Mountain, thus creating a protected area that included multiple forest systems under one project. And third, because human population densities are low around Crater Mountain, it was assumed that human-generated changes to the landscape are slight.

This chapter describes the complicated history of the CMWMA and the reasons for its decline. It is based on Paige West's research with the architects of the conservation-as-development project, the people who implemented it and carried out its day-to-day management, the scientists who conducted research within its boundaries, and its Gimi-speaking residents, as well as Enock Kale's experience as a facilitator working with Gimi communities and conservation organisations in an attempt in the early 2010s to try to revitalise the CMWMA.[1] The chapter also describes the decline of a program for training young Papua New Guinean scientists that grew out of the CMWMA and argues that even with the failure of the CMWMA this training program was an enormous conservation success for PNG.

Background to the Crater Mountain Wildlife Management Area

The CMWMA began informally with the work of David Gillison, a photographer and conservation enthusiast. David was married to anthropologist Gillian Gillison and in the 1970s and 1980s they lived with Gimi-speaking people in Ubaigubi village in Eastern Highlands Province. Gillian conducted anthropological field research concerned with gender and mythology and focused her work on Gimi women. David spent time with Gimi men, talking about initiation and trekking through the forests on hunting expeditions. During these talks and trips David 'fell in love' with New Guinea birds of paradise and began photographing them. He ultimately approached Wildlife Conservation International (what is now the Wildlife Conservation Society — WCS) with a plan to protect the habitat

---

1   Paige West is a cultural anthropologist who has worked in the Crater Mountain area since 1997 (see Ellis and West 2004; Mack and West 2005; West 2001, 2003, 2004, 2005a, 2005b, 2006a, 2006b, 2008a, 2008b; West and Brockington 2006; West and Carrier 2004; West et al. 2006). Enock Kale is a conservation biologist and a native Gimi speaker. He grew up in the CMWMA.

of these birds around four villages (Ubaigubi, Herowana, Maimafu and Haia) on a landscape that crossed the boundaries of Eastern Highlands Province, Gulf Province and Chimbu Province.

The CMWMA was officially established in 1994 (Johnson 1997: 397). In October of that year the PNG Department of Environment and Conservation declared it a national WMA under the *Fauna (Protection and Control) Act 1966*. The Act establishes the mechanisms by which 'Wildlife Management Areas, Sanctuaries, and Protected Areas' are set up and maintained. The Act provides for a set of formal institutionalised mechanisms to regulate wildlife harvesting, possession and trade in these areas.

For an area to be gazetted a national WMA, the 'customary landowners' must provide the Department of Environment and Conservation with a 'legal description of the boundaries of their area to be gazetted', a list of 'clan leaders who will sit on the local Wildlife Management Committees', and a list of conservation rules or laws that will be used to govern the WMA (*Fauna (Protection and Control) Act 1966*). Since the majority of people living around Crater Mountain had, at the time of the CMWMA's inception, little knowledge of the mechanisms by which the national laws work, WCS employee Jamie James went to the area to help people establish the boundaries and determine who would sit on the legally mandated Wildlife Management Committees (Johnson 1997: 399). James targeted 'leaders' who were hand picked by David Gillison, and assumed that because David identified them as 'leaders' that they were identified by the local communities as leaders. Most of the men identified as 'leaders' by Gillison had worked with him as field assistants during his years living in Gimi-speaking villages.

The Biodiversity Conservation Network (BCN) was the major donor for the Crater Mountain project during the period of its official inception. The largest single grant was the BCN implementation grant of US$498,107, which was spread over a three-year period from 1995 to 1998. As a BCN partner, the WCS committed US$76,950 in funding to the project.

BCN was designed to fund and study international conservation projects that linked biological conservation with economic development. The ideology behind BCN was that biodiversity as 'biological capital' is already linked to the economic 'health' of the planet.

The BCN program overview states:

> Conservation efforts that ignore the economic needs of local communities are unlikely to succeed. The Biodiversity Conservation Network (BCN), a component of the Biodiversity Support Program (BSP), is an innovative USAID-funded program working in the Asia/Pacific Region to provide grants

for community-based enterprises that directly depend on biodiversity. BCN is testing the hypothesis that if local communities receive sufficient benefits from a biodiversity-linked enterprise, then they will act to conserve it. (BSP 1996: iii)

BCN took as its premise that commodity production and economic incentives which tie people to commodity-based systems are the strategies that will promote the conservation of biological diversity. BCN devised what its literature refers to as the 'core hypothesis'. The BCN core hypothesis states:

If enterprise-oriented approaches to community-based conservation are going to be effective, they must: 1) have a direct link to biodiversity, 2) generate benefits, and 3) involve a community of stakeholders (BSP 1996: 1).

In effect, the hypothesis is that 'if local communities receive sufficient benefits from an enterprise that depends on biodiversity, then they will act to counter internal and external threats to that biodiversity' (BSP 1996: 1). The BCN project of hypothesis testing began in earnest with a US$20 million commitment from USAID in 1992 and was planned to last for six and a half years, ending in March 1999. Part of that money was used to found and fund the Research and Conservation Foundation of Papua New Guinea (RCF).

The RCF, a non-governmental environmental conservation organisation officially incorporated in 1986 with the help of the WCS and the PNG Department of Environment and Conservation, administered the Crater Mountain project. David Gillison founded the RCF because of his concern over what he thought to be 'declining populations' of birds of paradise in the forests near Crater Mountain. The stated goals of RCF with regard to the CMWMA fell directly in line with the stated goals of BCN. RCF attempted to meet these goals by implementing a series of programs within the four villages located in the WMA. These programs included creating local businesses revolving around biological research, tourism and handicraft production; teaching local men and children about biological diversity and conservation; and implementing a monitoring system to measure the results of biodiversity conservation.

In addition to all of this conservation-as-development around Crater Mountain, beginning in 1987 there were scientists working on the southern side of Crater Mountain. Americans Debra Wright and Andrew Mack began the research for their PhD degrees in ecology by building a research station that has become known as the Sera Research Station. The Sera Research Station was a biological research station located on land held by the Toyaido people of Haia (Pawaia people). It was built between 1987 and 1994 and served as the base for most of the biological research conducted in the CMWMA. Wright and Mack worked with Pawaia and Gimi people from almost the beginning of their research. Initially, they employed local people as carriers and guides but as the station came into being and their research projects got under way, they began to teach local men how to help them collect data. Because of this, in addition to the other 'income-generation' projects implemented by RCF, the architects of the CMWMA also incorporated some training of male landholders so that they could serve as research assistants, carriers and guides for scientists.

WCS PNG and the Development of Biological Science 'Training Courses'

Debra Wright and Andrew Mack served as the program directors for WCS PNG from 1999 to 2007.[2] Their original mandate was to conduct 'site based conservation' at Crater Mountain, manage and maintain the Sera Research Station, develop new sites for conservation and research in PNG, and conduct scientific research that could ultimately result in conservation in the country. In addition to these duties, Wright and Mack conceptualised, designed, developed and taught a series of biological science 'training courses' for PNG nationals. This was not part of their mandated duty for WCS; it was done in addition to their extensive job requirements.

The idea for the 'training courses' began to develop before Wright and Mack began working for WCS. In 1996, Mack (who was employed by Conservation International at the time) and Wright (who was still completing her PhD) applied for funds to conduct a three-month-long biological survey in three areas near Crater Mountain (Soobo, O-Pio and Aedo). They organised the survey around four taxa (birds, mammals, reptiles and amphibians, and plants) and took one University of Papua New Guinea (UPNG) student, one National Museum employee, one Forestry Research Institute professional, and one 'landowner' for each of the four taxa. During the survey they worked to 'train' their assistants in research methods including capture, identification and specimen preparation. During that same survey season, while at the camp they built at Soobo, they invited 20 students from UPNG to join them and take part in a course on methods. Both Mack and Wright enjoyed this course and the teaching that it entailed, but they were both struck by how little the senior students from UPNG knew about research design, methodology and data analysis.

2   Since 1999 WCS has maintained an office in Goroka, the city closest to the Sera Research Station at Crater Mountain. Over the years between the founding of the Goroka office and the end of the CMWMA, the WCS PNG office became the hub of New Guinea biological research. It hosted researchers from PNG, the United States, the United Kingdom, Australia, South Africa, Brazil, Germany, Venezuela, Mexico, France and Holland. All of these researchers went through Goroka on their way to the Sera Research Station, and while in Goroka they all stayed at one of the houses rented by WCS PNG, at 'Pacific Estates', a gated neighbourhood about one mile from the city centre.

Later that year Conservation International, with funding from USAID, asked Mack and Wright to conduct a rapid assessment program (RAP) survey at Lakekamu. They agreed under the condition that Conservation International fund a four-week visit from UPNG students during the course of the RAP. Conservation International agreed. Andrew Mack says:

> Debra and I took it in a slightly different direction in 1996 than the traditional RAP training as was done in Latin America. Our students came from a university with weaker fundamental skill building. Therefore, we incorporated more basic skills, like making data sheets, proper note taking ... We also made the project a bit more quantitative as this had been my objection to RAP prior to joining CI. I knew about [the Conservation International RAP method] from its inception because I had been a good friend of Ted Parker. But we [Debra and Andrew] both generally felt that RAP-style surveys could be enhanced without undue effort with a little bit of sampling rigor ... getting something more than just a species list.

In their attempt to refine and revise the RAP methods, Wright and Mack realised that with the program they were developing, they could pair the research they wanted to do to create a scientific basis for conservation in PNG with the biological science training of PNG nationals.

In 1997 they did another RAP-connected four-week course near Maimafu village, and in 1998 they did two more survey trips (again using the formula of one UPNG student, one National Museum employee, one Forestry Research Institute professional and one 'landowner' for each of the four taxa surveyed). Wright says:

> We began this massive training effort because students told us they didn't get any exposure to actual field research in their university training — certainly not to netting of birds and bats, trapping of mammals, etc. It was a need that we could fill. Without research skills biologists can't find out what they need to know in order to manage and sustain wildlife and ecosystems. We evolved the course each year based on post-course evaluations that asked the students what they liked and didn't like about the course, and what else they would like to see included. That's how we ended up doing alternate courses — one focused on field techniques, and one focused on project design and proposal writing. Both taught data analysis, report writing, and oral presentations. Each year we tried to hold the course in a different remote field site and we would take on between twenty-five and thirty UPNG students and conservation professionals for the four weeks. We thought it was important for us all to be living together out in the bush, as a team, working and learning together.

With these early surveys and RAP trips, Mack and Wright had been exposed to students from UPNG and felt they showed promise as researchers but needed to have some additional field-based training in methods and some additional laboratory-based training in analysis and statistics before they could work productively with visiting researchers or, more importantly, compete for international scholarships to study out of PNG for their graduate degrees. Because of this, they proposed to run a yearly course in research design, practices and analysis that would supplement the biological training students were receiving at UPNG. The course was aimed at the most capable and successful students majoring in the biological sciences and Wright, Mack with UPNG Professor Lance Hill choosing the students.

The training course ran for three years as a course only, but at the beginning of the fourth year Mack and Wright decided to add a second stage to the course. They, and Professor Hill, felt that some of the students who had taken part in the course had accomplished so much and learned so much in the course that they might be ready to conduct honours thesis work based on what they had learned. Wright wrote a grant proposal to fund a limited number of research projects to be designed and carried out by UPNG honours program students under the supervision of herself, Mack and Professor Hill. She was successful, and in 2000 they implemented a project that identified highly qualified graduates of their training courses, and offered them a position with WCS PNG as student researchers. In this position they would design a research project, carry it out, analyse the data collected and write an honours thesis. During this process they would be based in Goroka living in a house next door to the WCS PNG office and across the street from Wright and Mack.

Over the next six years a small campus developed around the WCS PNG office. We refer to it as a 'campus' because it became a kind of hub for researchers and students. Initially only people who were working at Crater Mountain came through the WCS PNG office, but by 2005 researchers who had once worked at Crater Mountain but who now had other sites in the country would stop by the WCS PNG offices and stay for a while, even if they were not going out to Crater Mountain. In addition, the Mack/Wright house, the home of the assistant director of WCS PNG and the senior national scientist, and the student dormitory house became lodging sites for visiting researchers from all disciplines. All of the researchers who moved through the WCS PNG campus were expected to spend time with the PNG students both in professional and social ways. The social world that emerged in Goroka was one that helped students develop both their research projects and their identities as scholars. It also served as the key node in the emerging network of scholars working in the CMWMA.

WCS PNG as a Conservation Success Story

WCS PNG, as it was organised under Wright and Mack and as it was connected to the CMWMA, was a 'success story' for conservation for four main reasons. First, Wright and Mack worked against the dominant model of expatriate scientist/national scientist interaction to build a program in which national scientists began to take the lead in scientific research design. Second, they fostered collaboration between international and national researchers from different disciplines. Third, they facilitated and conducted the most current and interesting scientific research in the highlands regions of the country. And fourth, they raised significant amounts of money for research and conservation in the country. That money did not go to consultant or expatriate salaries but rather to funding students from PNG to do research.

Historically there have been two forms of interaction between expatriate scientists and national scientists in PNG. The first, and most common, is what we term 'science as labour'.

Science as Labour

From 1995 to 2000 a UPNG-trained biologist lived and worked in Maimafu village, one of the villages in the CMWMA. During this five-year period he collected data on the biological diversity in the forests around Maimafu. From 2000 to 2002 a second biologist trained at UNITECH lived in Maimafu and collected data, from 2003 to 2004 another from UPNG and from 2004 to 2006 a fourth man lived there and collected biological data. None of these men understood fully why they were collecting this data. They knew how to do it (e.g., bird flyover counts, rain meter checks, wildlife trade counts) but they did not know what the data would be used for or what kinds of questions might be asked of the data. While these men were dedicated people, their lack of scientific knowledge of the very basics of rainforest ecology mean that as far as we are concerned the volumes of data collected are patchy and incomplete at best. But, in fact, American scientists have used some of those data in order to earn their PhD degrees. To date, none of the men who worked in Maimafu have used any of the data they collected for their own degrees. This mode of operation — 'science as labour' — is the mode of operation used in PNG by many of the major international conservation organisations working there. It does nothing to build capacity in country or conserve any biological diversity. It is one of the many reasons that 'BINGOs' (big international non-governmental organisations) have such a terrible reputation in PNG. It is, simply, another form of resource extraction. Like mining, logging and fishing, outsiders come in, extract what they want and leave little behind.

Papua New Guineans as Active Participants in Science

The second form of interaction, pioneered by anthropologist Ralph Bulmer and Ian Saem Majnep, treats Papua New Guinean nationals as active participants in the process of science, attributes scientific findings to them, and seeks to create avenues for them to become lead researchers and design research projects (Majnep and Bulmer 1977; Majnep and Bulmer 2007). WCS PNG fit this model during the Wright and Mack years. Instead of treating all people from PNG as labour, the organisation trained PNG nationals as scientists and local landholders as research assistants. Since 2001, Paige West has interviewed 26 nationals working in conservation in PNG for her ongoing research. These people work at various non-governmental organisations (NGOs) and government offices and 41 per cent of them attended one of the WCS training courses. With this, WCS PNG began to assure that the people moving up into the government and NGO offices of conservation in PNG have a solid background in their undergraduate training in the biological sciences. Because of financial constraints, the UPNG Biology Department cannot assure this.

Under the management of the Wright/Mack team, WCS PNG developed the aforementioned much-needed training programs for PNG nationals. These programs, both the biological field methods training course — in partnership with UPNG — and an additional teacher-training course that Wright developed in partnership with RCF and the University of Goroka, have contributed to the growth of national science and national science education in PNG. By developing these programs in partnership with PNG-based institutions the Wright/Mack team contributed not only to the growth of biological sciences in PNG but also to the growth of national institutional capacity.

The annual biological field methods training course, which began open only to UPNG students, now includes national participants from other conservation NGOs, national and provincial conservation-related departments, and both national museums. It has become the vanguard of science-based conservation training in PNG. Under Mack and Wright, WCS PNG also worked with many of the UPNG honours students to help them earn and win scholarships to study abroad for masters and PhD programs. In spring 2007 there were 12 honours students living in Goroka. To date, six of these students have gone on to masters programs in the United States, Australia and New Zealand. Previously, there had been ten successful honours degrees. All ten of these people went on to earn masters degrees. By 2011, the number had increased to 15, of which three were PhD candidates.

With all of this, WCS PNG built a network of national and international researchers that were connected through Andrew Mack and Debra Wright. The customary landholders around Crater Mountain came to recognise this network through their systems of meaning, making a patriclan (an extended family group based on patrilineal descent) with Andrew as the 'big man' and head of the family and Debra as his wife. Over the years, the Papua New Guinean students and the expatriate researchers who worked with WCS PNG became incorporated into social life around the Sera Station as Debra and Andrew's children, cousins and the like. Andrew was blamed when someone did something socially inappropriate and was lauded when someone did something that was seen as appropriate or good. These relations were instantiated through the long-term presence of some of the WCS PNG honours degree students in areas around the Sera Station and in the villages within the CMWMA. Mack was perceived as and acted as the patriclan leader and Wright was perceived as and acted as his wife. Mack came to embody the role of a highland big man in important ways and Wright came to embody the role of a mother.

None of this is to say that either Wright or Mack became, during their 20 years in PNG, fluent in or experts on local social articulations. Neither of them learned to speak the Pawaia language or the Gimi language. Neither of them actively engaged in learning the social vernaculars of the people they were working with in and around the Sera Station. In other words, they were one type of ecologist that we see globally — focused, when 'in the field', on their research, assuming that it had little or nothing to do with the local societies, so therefore less than interested in the local societies.

Resource Frontiers

While WCS PNG was working to establish its training program and to make the Sera Research Station the pre-eminent site for field-based scientific study in the country, various other international actors were also focusing on the Crater Mountain area. During the entire history of the CMWMA there has also been resource-related exploration in the area. The first mining exploration trip around Crater Mountain was in 1970, with subsequent trips in 1971, 1972, 1977, 1983, 1984, 1988, 1995, 1996 and 1997. These trips were undertaken by several different companies. All of the exploration trips after 1994 took place under the current exploration licence (EL 1115 Crater Mountain). This licence covers 'a rectangular area of 700 km between 6° 39' S 144° 56' E and 6° 28' S 145° 15' E and lies 50 km south-west of Goroka' with Crater Mountain sitting in the centre of the area (Macmin Silver Ltd 1997). With the issuing of this licence there was a revival in interest in the area around Crater Mountain and Macmin Silver Limited spent about US$1 million on a test drill site in 1998. After locating 'a major gold bearing hydrothermal system', Macmin considered it 'prudent to defer further valuable exploration funds' from Crater Mountain (Macmin Silver Ltd 2000). The reason given for this deferral was the low value of gold on the world market at that time, but the 'Crater deposit' was still, at that point, termed a 'company maker' for Macmin (Macmin Silver Ltd 2000). In 2002, New Guinea Gold Corporation, a partner company to Macmin, announced its acquisition of seven new gold and copper projects in PNG that included the Crater Mountain licence (NGG 2002).

Elsewhere, West has explained in detail what happened next:

> On 6 January 2004, a story in the Papua New Guinea *Post-Courier* reported that Macmin Silver Limited, a subsidiary of New Guinea Gold, had signed a joint venture agreement with Celtic Minerals Limited, which gave Celtic a 75 percent interest in the Crater Mountain project (*Post-Courier* 2004a). On 5 February 2004, New Guinea Gold issued a press release regarding the returns at Crater Mountain from a trenching program carried out in December 2003 (NGG 2004). The story was picked up by the *Post-Courier* and ran as a feature on the finance page on 9 February 2004 (*Post-Courier* 2004b). In the press release New Guinea Gold announced their application for an expanded exploration license they had filed with the PNG Department of Mining. According to the release they filed for this expanded license because Crater has such a favorable geological setting in terms of its potential for a 'major gold bearing mineralised system.' The *Post-Courier* story announced the trenching data (9.97 grams per tonne gold in one trench) and quoted Peter McNeil, the director of New Guinea Gold, as saying that they have 'never had results of this magnitude right at the surface.' The results indicated that this was a potentially lucrative new prospect, with an extremely rich vein, which the company continually likens to the Porgera area (Macmin Silver Ltd 2000). (West 2006b: 300)

After this frenzy of press releases, later in 2004 Triple Plate Junction, a gold and mineral exploration company from the United Kingdom, partnered with Macmin and Celtic and began to carry out further exploration in the Nevera Prospect Area and in early 2005 they began working in the Nimi Prospect Area. The land encompassed by the Nimi Prospect Area (the south end of EL 1115) is held in traditional tenure by people from Maimafu, according to people from Maimafu, and on land owned by people from Herowana, according to people from Herowana, and on land owned by people from Haia, according to people from Haia. These are three of the main villages that took part in the conservation-as-development project that was the CMWMA.

According to people from all of these villages, the prospectors who have visited their lands over the past 30 years have said, with each visit, that if a mine is established, people will be given roads, schools, hospitals, jobs and access to the other things that they see as development. Villagers also report that prospectors, geologists, gold buyers and the many people from other places who have flocked to the Crater Mountain area to attempt to access some of this mineral wealth say that if mining takes hold, development and cash will come quickly to the area.

In December 2003 InterOil, an oil exploration and refining company, landed at the Sera Research Station and told researchers and customary landholders that it would begin exploration activities in three days. InterOil said this would entail cutting at least five hectares of forest and bringing 200 workers to the site. WCS PNG attempted to negotiate with InterOil in order to have it move the exploration site out of the long-term study area (the area around the Sera Station had had few human impacts in terms of hunting or harvesting since about 1987). The company initially refused to negotiate with WCS PNG, citing underlying geological concerns as the reason for its decision to drill within the WCS PNG study area. Eventually, WCS PNG negotiated with InterOil a concession for minimised impact at the drill site and one for quick drilling (the company agreed to be out of the area by April 2004). After these negotiations, InterOil set up a 'camp' at the proposed drill site and staffed it with a small staff. Between December 2003 and March 2006, InterOil proceeded to cut seismic transects through the long-term study area surrounding the Sera Station but did not drill at all. According to WCS PNG staff, each time they contacted InterOil and asked about the drilling schedule, they were told that drilling would begin shortly.

The presence of InterOil near the Sera Station was not simple or benign. Local landholders began, soon after the camp was established, to visit and to sell food and other items to the small InterOil staff. Local Pawaia and Gimi peoples went to the camp to attempt to make money through these small commodity transactions and through gaining waged labour. They were told, repeatedly, that the staff would eventually hire them when the drilling began. The resident landholders were also told, according to people from Maimafu and Herowana villages, that one of the reasons for the delay in drilling, and thus the delay in waged labour being provided, was the presence of WCS PNG.

Conservation and mining offered the same things to landholders — they made resource extraction and conservation commensurable in people's eyes by tying them both to the promise of 'development' and the social relations that would bring development. But conservation-as-development as a possibility began to fade in local fantasies of the future by the mid-2000s. As WCS worked to establish its office and ongoing presence in PNG and as the resource extraction companies conducted exploration within the CMWMA, the RCF continued with its project of conservation-as-development even in the face of growing dissent among customary landholders who felt they were not getting what they had bargained for when they agreed to work with RCF initially.

The End of the Crater Mountain Wildlife Management Area

On 23 March 2006, four Gimi men from Herowana broke into the Sera Research Station with bush knives and guns. The station staff and various WCS PNG researchers and employees working at the station at the time were all in transition between Goroka, Haia (the village one flies into in order to then walk 12 hours to the station) and the station. There was a female researcher at the research station along with her field assistants who were all from Haia village. The four men from Herowana held a gun to her head, beat her and raped her. This vicious attack was the pivotal point in the history of the CMWMA. It was, more importantly, a horrific event for the victim. We write about it here so briefly because we are sensitive to this event not being ours to write about in any detail. However, this kind of 'punitive' rape is one increasingly common form of action in resource extraction situations in PNG. On 25 March 2006 WCS PNG pulled all of their staff out of the CMWMA.

Over the course of the months after the attack, Enock Kale met with people from Herowana to discuss the events that resulted in the violence of 23 March. Members of the Herowana community believed they were not benefiting from the conservation-as-development project. Their main complaint was that they felt that the profits from the project were not being distributed fairly. Initially, instead of blaming WCS, RCF or any of the researchers, they cast blame on Mr Avit Wako, a well-known community leader with historic ties to David Gillison. They argued that he was misusing the income generated from the project, unequally distributing the opportunity for villagers to participate as research assistants and guides, and selectively selling artefacts belonging only to his relatives to visitors. For instance, the villagers complained constantly about Mr Wako not disclosing the money paid as rent by WCS PNG in 2003, which was for bringing in and housing 30 students in the community guest house for one month during their annual biological training courses. The villagers estimated the total amount of money they lost during that training to be more than 10,000 kina. Additionally, many members of the community accused Mr Wako of nepotism with regard to deciding whose coffee bags would be the back load on the MAF (Mission Aviation Fellowship) planes that carried WCS's students and research supplies at a subsidised airfreight cost. The Herowana community viewed Mr Wako as a person who used his power to benefit himself or his relatives at the expense of the community.

Although much frustration was building up, many people could not speak against Mr Wako, because they saw him as intimidating; he would silence whomever he thought was speaking against him, with threats. WCS PNG, RCF and other visitors to Herowana did not find any fault in Mr Wako, because to them he was the best man to make logistics possible in the community. He could quickly organise carriers and field assistants for the WCS or RCF if needed. Therefore, according to the local community, he hid his bad deeds from WCS and RCF by promptly attending to their field logistics requirements. This was most likely why WCS was never aware of the complaints and frictions building up against Mr Wako by the Herowana community. These frustrations had even led some villagers to mobilise to construct another airstrip so that they can get away from Mr Wako's influence. The airstrip is almost completed to date.

While the complaints against Mr Wako were building up in Herowana, another land issue was developing between a Herowana family and the clan of Toyaido of the Haia area over the sharing of rent money that WCS was paying for the research site at Sera Research Station. The land on which Sera Research Station was built is rightfully owned by a clan in Toyaido village in the Haia area. It happened that one of the relatives of the clan that owned the land on which the research station was built had migrated to Herowana a long time ago. The descendants of this particular man in Herowana claimed that they were rightful landholders of the Sera Research Station and should therefore be receiving the rent money WCS was paying to the clan in Toyaido. The clan in Toyaido that was receiving the rent acknowledged the claim but provided one condition to the Herowana family before they would share the money: the Herowana man claiming to be the landowner must send one of his sons to be permanently settled in Toyaido with his related clan, and be completely a part of that clan, and then he would be the one with whom they would share the money. Otherwise, if he wanted to claim the money while in Herowana, then the money would not be paid to this man in Herowana or his sons.

The young men of Herowana began to observe the struggles between the community and Mr Wako's influence and decided to rob the Sera Research Station in the hope of disrupting the research activities from which Mr Wako seemed to be benefiting most. The four young men of Herowana who robbed the Sera Research Station and raped the female research scientist on 23 March 2006 were believed to be led by a man who is a direct descendant of the man who migrated to Herowana from the clan in Toyaido, who was also claiming to be the landowner of Sera Research Station but not receiving any rent.

In March 2007, Paige West, along with Robert Bino and John Ericho, the two former directors of the RCF (and both graduates of the WCS PNG training program and honours degree program), went to Maimafu village to hold a meeting with its residents to try to help RCF understand its failures.[3] When they arrived in Maimafu, they found no running water (the RCF-funded water supply system was no longer working), no school, no hospital, no guest house, no radio, limited ability to get coffee to markets, and a very angry mob. This was the first time that Robert and John had been to Maimafu in years. Paige West had been every year and was well aware of the decline there, but they were not. Although neither of them worked with RCF at the time (and indeed John had not worked for them since 2003 and Robert since 2005), people directly blamed John and Robert for the failures of RCF and the decline of the services that they saw as connected to RCF.

At the community meeting, several major concerns were expressed. First, the concern that RCF had 'gotten rich' off 'the Crater name' and that John and Robert had also 'gotten rich' and 'become big men' because of their work with RCF and 'with the Crater name' and that Maimafu's residents had 'gotten nothing' from their social relations with RCF in general and with Robert and John specifically. The second major concern was the failure of RCF to supply development or avenues for development and then totally pull out from the area with no explanation. This is basically the failure of RCF and its staff to participate in the correct way in the Gimi world of exchange relations. Third, people expressed their anger over the perceived opportunities they had missed because of their long-term association with the RCF. They referred to the mining opportunities they perceived missing and the oil-related exploration opportunities they perceived missing. One long-time RCF Management Committee member said:

> You told us that RCF would be a 'road to development' but what you have been for all these years now is a roadblock to development. You have stopped us and you have shamed yourself.

While landholders blamed individual actors for the failures of the project, it is our contention that the CMWMA was based on an untested and ultimately flawed hypothesis. Conservation-as-development, the driving ideological force behind the CMWMA, is a neoliberal approach to conservation ecology and economic development, in which it is assumed that environmental conservation can be economic development for rural peoples; that development needs, wants and desires can be met by the protection of 'biodiversity' on their lands, and that if they take part in small-scale income-generation projects that are directly

---

3   In 1994, while still a graduate student, Debra Wright asked WCS to fund a national student, John Ericho, during the research, analysis and writing of his honours thesis. Wright had met Ericho through then-UPNG professor Mike Hopkins. Ericho had been a high school biology teacher and principal at a Seventh-day Adventist school in Goroka for years and after meeting Mike Hopkins decided to go back to school to complete an honours degree. He was the first WCS-trained student who went on to be a director of the RCF. He is currently the national coordinator for the PNG Conservation Forum. Robert Bino worked in Maimafu village for the RCF as a village coordinator. He is currently completing his PhD in resource management at The Australian National University. He is not writing about the CMWMA.

connected to this 'biodiversity' they can seamlessly enter 'global markets' as producers and consumers, while at the same time working to conserve 'biodiversity' for the supposed good of the entire planet. These policies and practices, designed and implemented by NGOs, are meant to both conserve the natural world and provide cash income so that people living where state services have retreated can pay for basic needs like education, health care and subsistence. The projects also move the management and legislation of the relations between people, their surroundings and the market into the purview of NGOs. These eco-neoliberal practices and policies attempt to meet the social and economic needs of rural peoples through fostering, facilitating and supporting the retreat of the state and putting private industry and NGOs in its place. In the CMWMA the material and social consequences of these policies in and around rural villages have been that both people and the environment are left worse off as a result of their implementation.

Since the entire conservation area was based on the premise, from BCN, that conservation would bring people economic benefits, the project also, from its inception, was directly in competition with resource extraction. By connecting biodiversity to the global economic system, the architects of the CMWMA made conservation and mining commensurable. Residents of the CMWMA saw the two as equally valid options for economic and social betterment. They weighed their options, and chose the one that appeared to make the promise of more future benefits.

The CMWMA was also based on an erroneous set of assumptions about subjectivity that outsiders brought with them to their interactions with Gimi and the assumptions that Gimi had about outsiders. During the life of the CMWMA the incommensurable understandings of social relations and what those relationships meant held by conservation actors and Gimi peoples posed a barrier to true collaboration and understanding. For conservation planners and practitioners the conservation-as-development project established a contract between themselves and Gimi. In this contract, outsiders understood that they would get conservation, in the form of the curtailment of hunting practices and land clearing practices, and the allowing of access to Gimi territory in exchange for development, which took the form of cash income for Gimi based on ecotourism, handicraft sales and waged labour positions working for visiting scientists. Gimi, whose indigenous epistemology turns on the idea that all things come into being through exchange and that exchange relationships are in constant need of maintaining, understood the social relations they were entering into as enduring ones in which they would contribute land, labour, food and friendship to the conservation practitioners and that these contributions would be reciprocated in socially appropriate ways. They understood these social relationships as allowing them access to medicine, education, technology, knowledge and wealth — the things that they think of as 'development'. Through ethnographic evidence we demonstrate that this was not simply a set of misunderstandings but that these differences in the perception of reality illustrate different ways of being in and making the world. The conservation project's ultimate failure was not the result of people not holding up their end of the bargain; rather, people had been, for the duration of their social interactions, taking part in radically different social worlds.

These different social worlds also extend to the natural world or the Gimi forests. For Gimi, everything in the forest is a gift as it is the physical incarnation of their ancestors' life force. People and forests will always be — and have always been — in a constant transactive relationship, making and remaking each other over time. The Gimi world and Gimi as subjects and agents are produced through social relationships between people, ancestors, spirits and animals. These social relations are not neutral and economic; they are familial and poetic. People's capacities are seen as they relate to other people, to plants, to animals, to spirits and to ancestors, and their social selves are understood as comprising the sources that went into making them. So what is translated as 'the Gimi environment' by conservation practitioners is not simply a place filled with floral and faunal resources waiting to be used or made into commodities, it is the generative place of the social world. By treating Gimi forests as 'the environment', conservation practitioners not only misunderstood Gimi but also predestined their project for failure from the very beginning. The entire conservation-as-development project was based on ideas about nature and culture that define Gimi as rational economic actors who value forests because of their potential as commodities and status as resources. This assumption formed the basis for interventions — specifically ecotourism, handicraft production and attempts to regulate hunting practices — that have transformed society and space.

It was of no surprise to people familiar with the CMWMA that by 2006 RCF was seen as a total and complete failure in the eyes of the majority of landholders around Crater Mountain. In January 2006, RCF 'pulled out' of the Crater Mountain area completely. They flew their staff out of the villages without providing any explanations to the villagers and left without a trace. This quick exit was due, according to official emails and press releases, to the organisation's failure to raise funds for site-based conservation and development projects and its shift to primary and high school conservation-related education as a main focus.

The Fate of WCS in PNG and PNG in WCS

On Friday 9 March 2007, Paige West called Debra Wright to discuss an upcoming trip to PNG. When she answered the phone, Wright told West that WCS had terminated her and Mack's employment and that the fate of the WCS PNG program was completely up in the air. Over the course of the next six months various stories emerged about why Mack and Wright had been terminated.

The initial story given publicly about the Mack and Wright dismissal was that WCS had decided to 'pull out' of PNG totally. Various conservation-related actors were told that the country program was 'too expensive' and that WCS needed its limited resources elsewhere. When word began to spread that Mack and Wright had been fired, WCS began to receive outraged emails, faxes and phone calls from people all over the world. It received angry communications from well-known people: Sir Michael Somare (PNG's prime minister), Mal Smith (the governor of Eastern Highlands Province) and the Biology Department at UPNG. WCS also received hundreds of communications from scientists, social scientists and others who have worked in PNG over the years. The majority of these communications stressed the success of the Mack/Wright training program and voiced outrage over WCS's closing of its PNG office.

By mid-March 2007, the public story about the future of WCS PNG began to change. WCS International sent the assistant director of its Asia Program, a man who had never been to PNG before, and a WCS accountant to Goroka to meet with the WCS national staff to discuss 'the way forward'. These discussions resulted in Miriam Supuma and Banak Gamui, two WCS PNG employees from PNG who had been through the training program, the honours program, and successfully completed masters degrees (Supuma at James Cook University and Gamui at the University of Missouri), being named the 'co country directors' of WCS PNG. Supuma and Gamui were, over the next six months, slowly told that the education program at WCS PNG was to be phased out. The current students would be allowed to finish, but 'WCS is not a University' and Supuma and Gamui would be in charge of shifting the focus of the organisation to site-based conservation.

In an interview about the end of the WCS PNG training program, one very senior conservation biologist who has worked extensively in PNG told West:

> I would say any organization like WCS that has been in a country for twenty years has been there too long. They should be able to build the sort of national capacity needed in under twenty years if they really try, even in a place as backward (in terms of scientific capacity) as PNG. These big organizations, I think, see that attitude as some sort of threat to themselves. They don't really want to build a national capacity that will make them obsolete. I'm not sure I would call that

'neocolonialism' but maybe I would say they are subconsciously driven by unenlightened self-interest. After all these years of international conservation, WCS should have programs winking out around the world as national NGOs and government agencies take over. WCS could then concentrate on other countries and eventually, in an ideal world, be mostly focused on North America.

When they boast that they are committed to conservation in these places indefinitely, I see that as a flaw, not a worthy attribute. They should be boasting that they will not be needed indefinitely … that they will measure success not by how long they stay, but by how quickly they can leave.

Rebirth: The Papua New Guinea Institute for Biological Research

On 13 February 2008 the following notice appeared in both the *Post-Courier* and *The National*, the two major newspapers in PNG:

NOTICE OF INTENTION TO APPLY FOR THE INCORPORATION OF AN ASSOCIATION

I, Ms Miriam Supuma of Goroka, P.O. Box 277, EHP, a person authorized for the purpose by the committee of the association known as Papua New Guinea Institute of Biological Research, give notice that I intend to apply for the incorporation of the association under the *Associations Incorporation Act 1966*.

The following are the details of the prescribed qualifications for incorporation as specified in Section 2 of the Act:

(a) To be a premier research institution for conducting biological research and monitoring on the flora and fauna of Papua New Guinea; and

(b) To provide an avenue for national and overseas scientists to conduct biological research and monitoring to advance scientific knowledge in Papua New Guinea. We will do this conducting our own research and by aiding other scientists to conduct their research whenever possible, and by presenting our findings in an accessible manner to the scientific community, to the public, and to policy makers; and

(c) To provide a conducive learning environment for capacity building by training aspiring national biologists, including those currently attending universities as undergraduates or graduates and those already working as professional biologists and conservationists in the non-profit, government, or private sectors. We will do this by bringing in internationally recognized scientists to teach training courses and to give trainees first hand experience with biological surveys, research techniques, data analyses and scientific writing; and

(d) To source funding from national and international donor agencies to conduct biological research and monitoring, to train university students and conservation professionals to conduct biological research and monitoring, and to conduct seminars, education and awareness on biological knowledge and conservation in PNG; and

(e) To facilitate opportunities for university students and conservation professionals to attend short training courses, workshops and scientific meetings both in PNG and abroad and to further their careers through Honours, Masters and PhD degrees in PNG or at overseas institutions; and

(f) To collaborate with national and overseas universities, museums and other research institutions to conduct biological research in PNG and overseas and arrange for exchange programs to train PNG university students and conservation professionals; and

(g) To assist PNG university students and conservation professionals in terrestrial and marine biology, environmental science, forestry and agriculture to design and conduct appropriate biological research of high quality and of international standards; and

(h) To use our research data to develop recommendations for the sustainable management of PNG's fauna and flora and to present these recommendations to landowners, resources users, conservation managers, the government and other stakeholders to aid them in making wise decisions; and

(i) To build and maintain biological research station(s) where research will be done and collaborate with partner organizations on the use and management of the research station(s); and

(j) To produce and distribute papers, report and publications, media releases etc. on the research and work we do; and

(k) To plan and facilitate work both on advisory and consultancy basis; and

(l) To apply profits (if any) or any income in promoting the Association's objectives; and

(m) To prohibit the payment of dividends or payment in nature of dividend to its members; and

(n) To do generally all such other things that may appear to be incidental to or conducive to the attainment of the Association's objectives or any of the above.

After WCS International sent the assistant director of the Asia Program to PNG to sort out what to do with WCS PNG after the termination of Debra Wright and Andrew Mack, the senior Papua New Guinean staff of WCS PNG felt unsettled. In the initial discussions, even after two national ecologists (Miriam Supuma and Banak Gamui) were made co-directors of WCS PNG, WCS International argued that there had to be an expatriate advisor to the program. WCS International also argued that WCS PNG should be focused on site-based conservation and the creation of new protected areas. The senior staff of WCS PNG, all Papua New Guinean, felt strongly that the creation of new protected areas is not the appropriate strategy for fostering conservation in the country. They all saw the continuation of the training courses and the honours degree program as crucial for building a national conservation strategy. One of the staff, when asked why they all felt so strongly that the site-based conservation focus of the international program was a problem, said:

> I don't mean to disparage America, but protected areas is American and this is Papua New Guinea. With land ownership here, we can't alienate land. We need to figure out how to conserve our land ourselves and the way to do that is to produce more scientists from here. Scientists who know the place and the people and who know their science.

Because of this shared sentiment among the WCS PNG senior staff, they decided to form a new Papua New Guinean NGO and continue the training courses and honours degree program.

Conclusions

In this chapter we have described the history of the CMWMA and some of the reasons for its decline. We have also described one of the 'conservation success stories' to come out of the CMWMA and its decline. We showed that the project had very informal beginnings in the 1970s but that over the course of the 1980s and 1990s its architects worked to solidify it spatially and legally under Papua New Guinean law. We then showed that in the late 1990s the project was incorporated into a large international 'experiment' meant to test the hypothesis that market or enterprise-oriented approaches to conservation are effective if indigenous peoples are given labour connected to conservation and if they are encouraged to commoditise biological diversity in particular kinds of ways. Next we described in detail a program for science training and education for national scientists that grew out of the CMWMA. Finally, we discussed the demise of the CMWMA and of the training program, highlighting the material and ideological reasons for these declines. It is our contention that the CMWMA was based on a fatally flawed set of ideas and that it was carried out as if larger political–economic drivers were not conterminously affecting landholder life and politics as the CMWMA was being developed. The living legacy today of the CMWMA is the PNG Institute for Biological Research. This NGO has facilitated the growth of national science in the country and created a generation of national scientists and conservationists. It is also our contention that conservation projects developed by these actors will outperform interventions like the CMWMA.

References

BSP (Biodiversity Support Program), 1996. *Biodiversity Conservation Network 1996 Annual Report: Stories from the Field and Lessons Learned.* Washington: BSP.

Ellis, D.M., 2002. Between Custom and Biodiversity: Local Histories and Market-Based Conservation in the Pio-Tura Region of Papua New Guinea. Canterbury: University of Kent, Department of Anthropology (PhD thesis).

Ellis, D.M. and P. West, 2004. 'Local History as "Indigenous Knowledge": Aeroplanes, Conservation and Development in Haia and Maimafu, Papua New Guinea.' In A. Bicker, P. Sillitoe and J. Pottier (eds), *Investigating Local Knowledge: New Directions, New Approaches.* London: Ashgate Publishing.

Johnson, A., 1997. 'Processes for Effecting Community Participation in the Establishment of Protected Areas: A Case Study of the Crater Mountain Wildlife Management Area.' In C. Filer (ed.), *The Political Economy of Forest Management in Papua New Guinea.* Boroko: National Research Institute (Monograph 32).

Mack, A. and P. West, 2005. 'Ten Thousand Tonnes of Small Animals: Wildlife Consumption in Papua New Guinea, a Vital Resource in Need of Management.' Canberra: The Australian National University, Resource Management in Asia-Pacific Program (Working Paper 61).

Macmin Silver Ltd, 1997. 'E.L.1115 Crater Mountain Site Report.' Unpublished manuscript.

Macmin Silver Ltd, 2000. 'Macmin N.L. Current Projects Crater Mountain.' News Release, 9 February.

Majnep, I.S. and R. Bulmer, 1977. *Birds of My Kalam Country.* Auckland: Auckland University Press.

Majnep, I.S. and R. Bulmer, 2007. Introduction. In I.S. Majnep, R. Bulmer, R. Hide and A. Pawley (eds), *Animals the Ancestors Hunted: An Account of the Wild Mammals of the Kalam Area, Papua New Guinea.* Adelaide: Crawford House Publishing.

NGG (New Guinea Gold), 2002. 'New Guinea Gold Corporation (the Company) to Acquire up to 100% Ownership of Seven Papua New Guinea Gold & Gold/Copper Projects.' News Release, 12 June.

NGG (New Guinea Gold), 2004. 'Crater Mountain Trenching Returns High Gold Results.' News Release, 5 February.

Post-Courier, 2004a. 'Macmin Silver Ltd. Signs Joint Venture Agreement with Celtic Minerals Ltd.' *Post-Courier*, 6 January.

Post-Courier, 2004b. 'High Gold Results at Crater Mountain.' *Post-Courier*, 9 February.

West, P., 2001. 'Environmental Non-Governmental Organizations and the Nature of Ethnographic Inquiry.' *Social Analysis* 45(2): 55–77.

West, P., 2003. 'Knowing the Fight: The Politics of Conservation in Papua New Guinea.' *Anthropology in Action: Journal for Applied Anthropology in Policy and Practice* 10(2): 38–45.

West, P., 2004. 'Environmental Non-Governmental Organizations and the Nature of Ethnographic Inquiry.' In P. Stewart and A. Strathern (eds), *Anthropology and Consultancy: Issues and Debates*. New York and Oxford: Berghahn Books.

West, P., 2005a. 'Holding the Story Forever: The Aesthetics of Ethnographic Labor.' *Anthropological Forum* 15(3): 267–75.

West, P., 2005b. 'Translation, Value, and Space: Theorizing an Ethnographic and Engaged Environmental Anthropology.' *American Anthropologist* 107(4): 632–42.

West, P., 2006a. *Conservation Is Our Government Now: The Politics of Ecology in Papua New Guinea*. Durham (NC): Duke University Press.

West, P., 2006b. 'Environmental Conservation and Mining: Between Experience and Expectation in the Eastern Highlands of Papua New Guinea.' *The Contemporary Pacific* 18(2): 295–313.

West, P., 2008a. 'Conservation Actions and Events in Papua New Guinea.' In B.B. Walters, B.J. McCay, P. West and S. Lees (eds), *Against the Grain: The Vayda Tradition in Human Ecology and Ecological Anthropology*. Lanham (MD): AltaMira Press.

West, P., 2008b. 'Scientific Tourism: Imagining, Experiencing, and Portraying Environment and Society in Papua New Guinea.' *Current Anthropology* 49(4): 597–626.

West, P. and D. Brockington, 2006. 'An Anthropological Perspective on Some Unexpected Consequences of Protected Areas.' *Conservation Biology* 20(3): 609–16.

West, P. and J.G. Carrier, 2004. 'Ecotourism and Authenticity: Getting Away from It All?' *Current Anthropology* 45(4): 483–98.

West, P., J. Igoe and D. Brockington, 2006. 'Parks and Peoples: The Social Impact of Protected Areas.' *Annual Review of Anthropology* 35: 251–77.

8. How April Salumei Became the REDD Queen

COLIN FILER

Introduction

The Papua New Guinea Forest Authority (PNGFA) maintains a database containing all of the forest areas that have ever been designated as potential logging concessions by means of agreements between their customary owners and the national government or the former colonial administration. In a version of this database that dates to the end of 2011, there are 314 such areas, covering a total of 10,953,897 hectares, which is almost one quarter of PNG's total land area. Most of the agreements had 'expired', which means that the areas in question had almost certainly been logged at some time in the previous 50 years; many were 'current', which means that logging operations were probably ongoing; and a few were under dispute. But one area stands out from all the rest. The 521,500 hectares that comprise the April Salumei forest area in East Sepik Province are the site of the only 'REDD+ Pilot Project' (GPNG 2012a, Appendix 4). This chapter will seek to explain how this particular forest area came to acquire this special status.

Google will tell you many things about April Salumei, but one of the more curious things can be found on a website called 'The Carbon Capture Report', which has a 'profile page' that 'demarks this person as appearing in content suggesting contextual association with Carbon Credits'. The content in question consists of a single post made in August 2011 by a PNG blogger who goes by the name of Masalai — the Tok Pisin term for a bush spirit. The Bush Spirit wondered if the carbon tax which had just been introduced by the Australian government would act as a 'catalyst for carbon cowboys' to repeat a form of engagement with PNG's 'carbon credits' which had been the subject of so much controversy back in 2009. Specific reference is made to an Australian company called Carbon Planet as 'the company that would buy and trade the carbon credits that Kirk Roberts was busy quantifying in the forests of April Salumei' (Anon. 2011). It sounds as if poor April may have been an April fool to let that particular 'carbon cowboy' count her carbon credits. And yet she seems to have survived and prospered in the safer hands of government officials.

The Bush Spirit was wrong to suggest that Carbon Planet and Kirk Roberts ever had much to do with April Salumei, but he was right to say that all three of them had figured in a sort of moral panic that enveloped the forest carbon policy process in 2009 (Pearse 2012; see also Wood, Chapter 9, this volume). The origins of this moral panic can be traced back to a political contest that started in 2005, when the prime minister and self-styled grand chief, Sir Michael Somare, made PNG one of the founding members of the Coalition for Rainforest Nations. The aim of this body was to amend the United Nations (UN) Framework Convention on Climate Change in order to enable countries like PNG to receive compensation from the international community for action taken to reduce greenhouse gas emissions from deforestation and forest degradation (REDD). A proposal along these lines was accepted in principle at the UN 11th Conference of the Parties in December 2005 and has been subject to further negotiation ever since. However, the prospect of REDD compensation payments was not greeted with universal enthusiasm in PNG. That was largely because the Coalition for Rainforest Nations was known to be the brainchild of Kevin Conrad, an American friend of the Somare family who had recently acquired a degree in business administration from Columbia University, and whose previous business ventures in PNG had been somewhat controversial (Lang 2010a).

For a brief period in the middle of 2005, there was an intense but inconclusive public debate about the relationship between Conrad's terms of engagement as PNG's 'climate change ambassador' and the apparent transformation of national forest policy (Filer and Wood 2012: 668). A sense of popular confusion and suspicion in the wake of this debate was expressed in a letter written to one of the national newspapers:

> Supposedly carbon swaps means money for the landowners to protect their forest. In reality, our government will give the same ridiculous small percentage to the landowners that they do with any other resource development. The Government will walk away with nearly all the revenue and probably the land rights of the forests ... Next time you see someone promoting carbon swaps, look behind the scenes. Smell the money and find out where most of it goes. (Bamu 2005)

Although this public debate did not last long, it did foreshadow the more intense debate that came to a head in 2009, in which foreign 'carbon cowboys' were accused of fomenting a 'carbon cargo cult'. Some of the people making these accusations thought Conrad was wholly or partly to blame for this moment of irrational exuberance, yet Conrad deplored it as loudly as they did. As a result, the national government took strong measures to suppress the voluntary carbon schemes which the so-called cowboys were promoting (Filer and Wood 2012: 669–70).

Nearly all of the voluntary schemes have since been thrown into the dustbin of economic history. April Salumei seems to have escaped this fate because she got a stamp of government approval. However, the moment of irrational exuberance was partly fostered by the distribution of these very same stamps to many other projects that have not survived the subsequent rationalisation of forest carbon policy. So there is still a mystery. My attempt to solve this mystery will proceed as follows: in the next section, I show how the idea of a national 'carbon cargo cult' came to be associated with representations of the April Salumei forest area. In the following section, I leave that discourse to one side and recover what is known of negotiations over the development of this area before it came to the attention of the cult's opponents. In the remainder of the chapter, I analyse the social and political process by which the April Salumei REDD project[1] has achieved a distinctive form of validation at both national and international scales. I conclude with a brief discussion of the reasons for this apparent success.

The Carbon 'Cargo Cult'

In September 2009, the *Weekend Australian* newspaper published a feature article by Asia-Pacific editor Rowan Callick under the title: 'The Rush is on for Sky Money: The Struggle Begins to Control the Immense Wealth Certain to Flow from Carbon Trading'. According to Callick (2009), the carbon trade in PNG 'has been converted into a cargo cult luring cash from Australian taxpayers, international investors and local villagers alike'. The Tok Pisin phrase *mani bilong skai* ('sky money') was said to have been coined as a reference to the profits to be made from the sale of carbon emissions, and reference to a 'cargo cult' was partly justified by evidence that some local villagers had already given real money to 'local confidence tricksters [who] have been trawling through rural

---

1   Although it has been described as a 'REDD+' project, the plus sign is superfluous. The 'plus' was added to REDD at the Copenhagen climate change conference in 2009 to indicate the possibility of securing carbon credits by making forest management more 'sustainable' or planting new forests (see Gabriel, Chapter 10, this volume). The April Salumei project conforms to the older definition of a REDD project as one that simply aims to prevent deforestation and/or forest degradation.

areas, urging people to register with their fake firms to make millions of kina from trading to developed countries the carbon values embedded in their trees' (ibid.). The main source for the whole story was Ilya Gridneff, the Australian Associated Press correspondent in Port Moresby, who was listed as the second author when the story was reproduced in the *Post-Courier* newspaper (Callick and Gridneff 2009). Gridneff had already published an article in *The National* newspaper in June 2009 in which he cited David Melick of the World Wide Fund for Nature (WWF) as his own source of information for the popular belief that 'sky money' could be obtained by capturing carbon emissions and bringing them for sale in the national capital (Gridneff 2009a). The implication was that villagers had got the wrong end of the stick when hearing someone talk about technologies for carbon capture and storage.

In June 2008, David Melick was already writing about 'carbon cowboys' as '[u]nscrupulous carbon traders (linked to highly placed political figures)' who were 'strong-arming and threatening landowners' in a manner already familiar in a national forestry sector 'where corruption is rife' (Melick 2008). However, a full year passed before the internet was suddenly abuzz with evidence of strange dealings between some of these carbon traders and Theo Yasause, then executive director of PNG's Office of Climate Change (OCC).[2] In November 2008, an editorial in the *Post-Courier* warned of a possible invasion of 'speculators' and 'carpetbaggers' looking to make a fast buck out of the carbon business (Anon. 2008), but most of the local news in 2008 was about the role of the national government and other stakeholders in the design of an institutional framework that would enable respectable foreign investors (including the governments of countries like Australia and Norway) to pay for the conservation of forest carbon. That sort of news did not make for sexy headlines around the world. The media feeding frenzy only started in June 2009, with a stream of articles and blogposts from a number of journalists — most notably Natasha Loder, an American journalist who was writing articles for *The Economist* magazine, and the aforementioned Ilya Gridneff, who developed a close working relationship with Loder from his base in Port Moresby. In November 2009, Loder and Gridneff were jointly awarded a prize for their efforts by the UN Correspondents Association, but they had run out of scandalous revelations by the end of September, so the story of PNG's 'carbon cowboys' was almost entirely assembled and disseminated over a period of four months.

What the journalists discovered in June 2009 was a set of documents which appeared to show that Theo Yasause had been granting or selling rights to trade in the carbon contained in various forested areas, facilitating deals between

---

2   This body was initially known as the Office of Climate Change and Carbon Trade, then as the Office of Climate Change and Environmental Sustainability, and more recently as the Office of Climate Change and Development. In this chapter I refer to all its incarnations by the abbreviated title.

carbon trading companies of different size and provenance, or endorsing the allocation of carbon trading rights that had been made by a government minister back in 2005 — more than three years before the OCC was formally established in September 2008. In July 2009, leader of the opposition Mekere Morauta made a speech to parliament in which he quoted from a letter supposedly written by then minister for trade and industry, Paul Tiensten, to a company called Climate Assist (PNG) Ltd in October 2005, appointing that company as 'brokers on behalf of the Independent State of PNG to buy and sell carbon credits' (Morauta 2009). Morauta would later say that he had come across another letter, sent from the acting secretary for trade and industry to the chief executive director of the same company in June 2005, which confirmed the national government's acceptance of a payment of US$200 million in return for '33,333,333 metric tones' [sic] of carbon credits to be traded in accordance with various provisions of the Kyoto Protocol (Talu 2009). In his parliamentary statement, Morauta said that a search of Australian company records had revealed that Climate Assist (PNG) Ltd was 'a one dollar company' based in Queensland, which had since entered into an arrangement with another company called Earth Sky Ltd, registered in the British Virgin Islands, to advance A$10 million to the OCC 'in return for the rights to sell $500 million carbon offsets, retaining 20% for themselves'. He also wondered how the OCC had 'authorised a Swiss based broker, South Pole Carbon Asset Management to market 1 million tonnes of avoided carbon dioxide emission[s] per annum from a PNG logging project based in the Sepik, April Salome'.

Some journalists had already made a link between April Salumei, South Pole and Earth Sky (Wynn and Creagh 2009), and another opposition member of parliament (MP) had called for an investigation of the deal between OCC, Earth Sky and Climate Assist at the very start of the media feeding frenzy (Anon. 2009a). The link between all five entities was present in the first draft of a design document, produced in December 2008, for the April Salome Sustainable Forest Management Project (ASSFMP). This contained a copy of a memorandum from Theo Yasause to the prime minister, apparently written in June 2008, on the subject of 'Collaboration with Earth Sky and Climate Assist PNG'. According to this document, the two companies had 'already made representation to the National Forest Authority, the Department of Trade & Industry, National Planning and various land groups in Central, Madang, Morobe, Milne Bay, New Ireland, Eastern Highlands, Western Highlands and Southern Highlands'. It went on to say that they would now be 'remitting close to US$50 million per month to the Office of Climate Change to cover for projects in the country for the preservation of loggable forest' in return for 'a certificate to go out and sell the credits' (Earthsky 2008: 31). As part of this deal, the two companies would retain 5 per cent of the revenues from such sales, the OCC would retain 20 per cent, 20 per cent would be paid directly to landowners, 15 per cent would be spent on 'community projects', 20 per cent would be placed in a 'future generation fund', and 20 per cent would be turned into a bond 'to cover for forest degradation'. Yasause estimated that REDD projects in eight provinces would 'account for 33 million metric tones of carbon credit', which 'could bring to the country close to $500 million US dollars over the next three years'. In order to set this ball rolling, the 'above group' was apparently prepared to invest A$10 million in the establishment of the OCC itself (Lang 2009a; Leggett and Lovell 2012: 122).

The mention of '33 million metric tones' — which one imagines to mean tonnes — can hardly be a coincidental echo of the earlier letter mentioned in Morauta's speech to parliament. But the link to April Salumei (or Salome) is rather more tenuous, because East Sepik Province is not one of the eight provinces from which this weight of carbon credits was meant to be derived. The link was only made by inclusion of a copy of the memorandum in the project design document which named Earthsky Ltd as the 'owner' of the project and South Pole as the 'carbon asset manager' (Earthsky 2008: 15). In his own response to the glare of negative publicity which enveloped his office in June 2009, Yasause declared that he had stopped dealing with Climate Assist in September 2008, but the prime minister, the forests minister and the local MP 'have been kept fully informed and involved and are fully supportive of the April Salome Project' (Anon. 2009b). The other potential supporter he failed to mention was Kevin Conrad, who had taken a high-powered Norwegian government delegation on a visit to April Salumei in May 2008.

The glare of negative publicity was not the coincidental result of some brilliant homework by a small group of investigative journalists. The diet of dodgy documents that fed the media feeding frenzy was served up by a group of national political actors who thought that a well-aimed attack on Theo Yasause would bring down the whole house of cards constructed by Kevin Conrad (Lang 2009b). But they were disappointed. Yasause was suspended and later dismissed, but Conrad's grip on forest carbon policy was not relaxed as a result. He soon announced his own determination to rein in the 'carbon cowboys ... [who were] confusing forest communities with tall tales of gold trains just around the corner' (Gridneff 2009b). Whatever he had in mind when he showed April Salumei to the Norwegian government, his new campaign against the carbon cowboys was justified by his previous insistence on the need for a 'national' approach to REDD, not for the state-sponsored 'project-based' approach authorised by Yasause or the purely voluntary version of that approach favoured by some of the non-governmental conservation organisations (Howes 2009; Filer and Wood 2012). This was made evident in a media release that Conrad most likely wrote for his patron, the prime minister, in which it was announced that 'my Government does not currently see a role for Voluntary Carbon Agreements in its policy development and regulatory framework for forest carbon and climate change in general' (Somare 2009). The secretary for environment and conservation, Wari Iamo, who had been put in charge of the OCC while Yasause's actions were being officially investigated, later published another public notice pointing out that '[c]arbon trading agreements cannot legally be signed' over land subject to a Forest Management Agreement (FMA) 'until the Government has put in place an appropriate policy and legal framework' (OCCES 2009). The moment of irrational exuberance was officially dead.

April in Waiting

According to one estimate, more than 90 carbon trading schemes were promoted during that brief period of history, with a combined coverage of more than 5 million hectares of forested land (Melick 2010: 360). Most of the schemes have never been identified in the mass media or in the substrate of grey literature from which the journalists drew most of their inspiration. April Salumei was a very large area of land containing a large amount of unlogged forest, but what really set her apart from the crowd was not so much her physical attributes as her peculiar political history. The PNGFA thought it had acquired the timber harvesting rights from the customary owners by means of an FMA, but the status of this transaction was shrouded in uncertainty. She looked like a REDD project in waiting because forestry officials could not think of a way to turn her into a normal logging concession (Filer 2012: 612).

In March 1992, the former Department of Forests signed a Timber Rights Purchase agreement over the April Salumei forest area with 116 'clan agents' appointed to represent the customary landowners, but 41 of the signatories represented Sepik River communities outside the prospective logging concession 'who were supposed to share benefits because of "disturbance" of the river by logging activities' (GPNG 2001: 6). Whether for this or other reasons, the validity of the agreement was challenged in court, so the National Forest Board decided to negotiate a new FMA with the customary owners under the provisions of the new Forestry Act, which had come into effect in June 1992 and turned the Department of Forests into the PNGFA. Following the provisions of the new legislation, officers of the National Forest Service then arranged for the incorporation of 164 (or possibly 181) land groups to represent the customary landowners, and in December 1996 the chairmen of 129 (or possibly 133) of these land groups signed an FMA with the PNGFA (GPNG 2001: 6, 2009: 4028; Leggett 2009: 45).

In the meantime, a consortium of environmental non-governmental organisations (NGOs) known as Friends of the Sepik had been working with the East Sepik Council of Women and a number of local villagers to oppose the prospect of a large-scale logging operation and opt for an integrated conservation and development project instead (Duguman 2004: 3). In 1995, the World Wide Fund for Nature (WWF) took the lead in helping the dissident landowners to get their land excluded from the FMA area and persuade the Department of Environment and Conservation to recognise it as a Wildlife Management Area (WMA) under the provisions of the Fauna (Protection and Control) Act. An area of roughly 230,000 hectares was duly gazetted as the Hunstein Range WMA in November 1997 (Leggett 2009: 46). Some parts of the new WMA lay outside the boundaries of the April Salumei forest area, but an area of roughly 180,000 hectares was covered by both agreements (Figure 8.1).

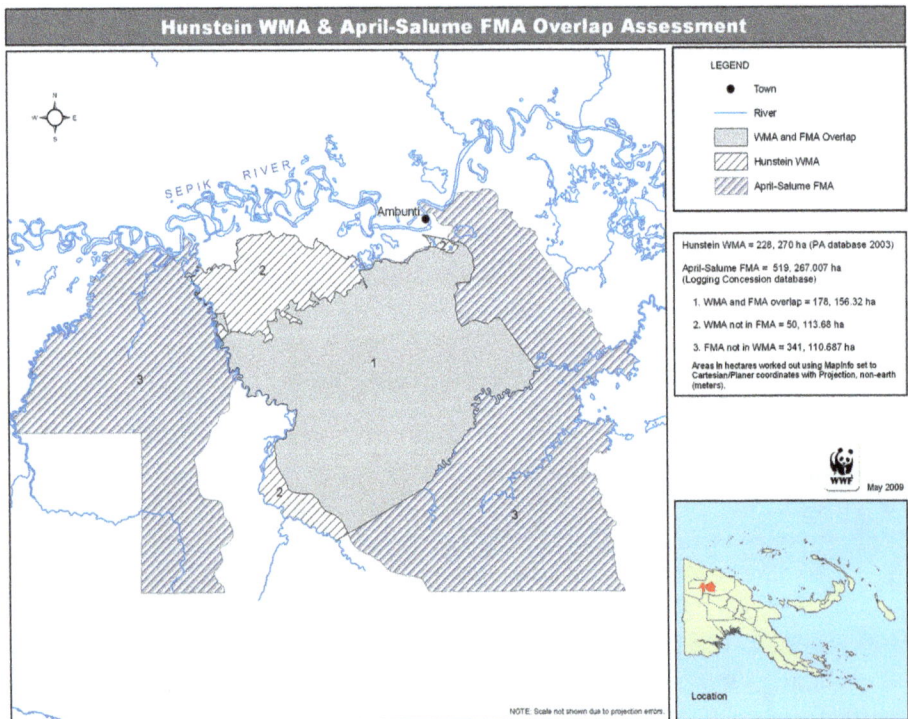

Figure 8.1 The spaces at stake in the April Salumei area.

Source: Map courtesy of WWF, Western Melanesia Program.

This now became a test case in the relationship between logging concessions and protected areas, and also a test of the relationship between the two government agencies responsible for such things. In an unprecedented gesture of goodwill, the PNGFA 'gave' K40,000 to the Department of Environment and Conservation to conduct an environmental assessment of the FMA area (GPNG 2009: 4025), and this was duly presented to the National Forest Board in February 1999, with the observation that 44 per cent of the forest was 'fragile', and a polite recommendation that logging should be avoided if possible (GPNG 2001: 3). The board was not legally obliged to accept this recommendation, since the declaration of a WMA does not transfer any property rights from customary landowners to the state, but since the PNGFA already knew that a number of landowner representatives had not given their consent to the state's acquisition of the timber harvesting rights, nothing more was done to advance the process of 'resource allocation' prescribed by the Forestry Act. In 1998, WWF secured funding from the Netherlands government for the Sepik Community Land Care Project, with a range of activities directed towards the conservation and development of the Hunstein Range WMA. This project should have lasted for six years, but the funding was unexpectedly withdrawn after five years and the project's capital assets were donated to the East Sepik provincial administration (Duguman 2004).

All of the external agencies and organisations that have sought to establish some sort of claim over the resources of the April Salumei area have been baffled by its cultural diversity and political complexity. None of the interested parties has so far been able to produce a definitive list of the villages or census units that contain the local 'landowners', but they all seem to belong to one or other of the local-level government (LLG) council wards whose populations are shown in Table 8.1. Each ward contains between one and three census units.[3] The residents of these 27 wards speak 13 or 14 languages, most of which were not traditionally spoken outside the area. Most of the village settlements in the area contain less than 200 people, and these local communities were traditionally at war with at least half of the neighbouring communities, even with those that spoke the same language (Leggett 2009: 63). Nor have the institutions of the modern state done much to enlarge the scale of cooperation between them. Different parts of the FMA area overlap the areas represented by three elected members of the national parliament, as well as the four local councils, but 'April Salumei' people constitute a small minority of the people represented in each electorate, and most candidates would not even bother to canvass their votes.

3   Leggett and Lovell (2012: 123) have published a map showing the local government boundaries and distribution of census units in the area covered by Figure 8.1, but the level of detail makes it hard to decipher.

Table 8.1 Populations of council wards containing customary owners of the April Salumei forest area in 2000.

Tunap-Hunstein LLG area		Ambunti LLG area	
Walio	132	Kupkain	318
Nein	117	Prukunawi	243
Wusok	29	Malu	661
Wasuware	25	Yerakai	432
Sio	200	Garamambu	621
Hanasi	166	**Subtotal**	**2275**
Moropote	298		
Maposi	201	**Gawi LLG area**	
Lariaso	66	Changriman	208
Yabatawe	144	Mari	163
Sowano	252	Yembiyembi	233
Bitara	192	**Subtotal**	**604**
Kagiru	161		
Begapuki	179	**Karawari LLG area**	
Gahom	134	Meska	443
Wagu	474	Bisorio	316
Niksek	489	**Subtotal**	**759**
Subtotal	**3259**		
Total population of wards in all four LLG areas			**6897**

Source: PNG national census data.

The political fragmentation of the area has been reflected in the multiplication of bodies claiming to represent the rights and interests of the customary landowners. A survey of six villages in the area indicated the presence of 53 'principal clans', of which only 24 had been given legal recognition as incorporated land groups (ILGs) (Leggett 2009: 62), but anthropologists who have worked in this part of the country doubt whether 'clans' actually function as the holders of collective property rights in any meaningful sense (Guddemi 1997). At the time when the FMA was being negotiated with the leaders of those 'clans' that had been incorporated, there were apparently three landowner companies claiming to represent the interests of those landowners who preferred logging to conservation, but shortly after the agreement had been signed, five such companies apparently agreed to form an 'umbrella company' called April Salumei Resource Development Corporation Ltd (GPNG 2001). It is not clear what purpose any of these companies was meant to serve, since the Forestry Act does not recognise landowner companies as legitimate parties to an FMA. On the other hand, some of the land group chairmen who signed over 'their' timber harvesting rights to the state in 1996 were among the landowner representatives who teamed up with members of the April-Salumei Landowners Association to support the declaration of a protected area one year later (Leggett 2009: 46). A subsequent review of the FMA summed up the state of confusion as follows:

> It is not clear as to which ILGs 'belong' to which LANCO [landowner company]; which ILGs have signed up for forestry and which are working with WWF on non-logging use of the forests; did the April people stay out of the FMA for ethnic or conservation reasons; which ILGs that have signed the FMA are now wanting to withdraw; are the forest owning ILGs happy to have the river ILGs in the FMA since the latter own no forest? (GPNG 2001: 8)

This also helps to explain the National Forest Board's decision not to tender the concession.

This decision was subsequently contested by Hunstein Range Holdings Ltd (HRH), which was one of the landowner companies involved in the previous rounds of negotiation. Jackson Yagi, the chairman of this company, hails from Yerakai village in the north-eastern corner of the FMA, beyond the border of the WMA which sat on top of it (Figure 8.1). Yagi has claimed that HRH was first incorporated in 1995, and that its shareholders were the 129 land groups whose chairmen (including himself) agreed to sign the FMA in 1996 (GPNG 2009: 4033). This claim seems to be at odds with PNGFA records, but those records do suggest that HRH was promoting the interests of one particular logging company as the 'preferred developer' of the area (GPNG 2001). When the National Forest Board expressed its own preference for the area not to be logged at all, HRH sued the state for the costs it had incurred in its own private dealings.

The outcome of this action would still be shrouded in mystery if it were not for the fact that Yagi and two of his legal advisers later testified before a commission of inquiry investigating allegations of malfeasance in PNG's Department of Finance. According to this testimony, the state agreed to pay K930,100 in compensation to HRH, and K430,000 was in fact paid out in 2005, while another cheque for K380,000, originally issued in April 2008, had 'gone stale' by July 2009 because the solicitor general refused to approve the payment until the commission dealt with the case (GPNG 2009: 4041). In response to questions about the basis of his claim, Yagi said that his board had incurred a lot of travel and accommodation expenses for which it had been obliged to take out something 'like a loan' for 'just over a million kina' from a Taiwanese company called Raw Timber Ltd (ibid.: 4038). He then went on to say that the money received to date had not been used to repay this debt because HRH had incurred more expenses on its own account (ibid.: 4048), and 'today I have taken out another order which the National Court actually granted which was in favour of my landowner company for the entire contract agreement so it has gone up to some 60 million' (ibid.: 4050). However, in a newspaper article published on the day of the hearing, Yagi was reported as saying that he wanted the commission to 'fast track' the payment of money under the previous settlement 'to save the State paying K50.8 million as benefits estimated under the Forest Management Agreement' (Anon. 2009c).

Jackson Yagi's interest in the opportunity costs of forest conservation had already been matched by a parallel interest in its potential benefits. In January 2004, he was congratulated by his local MP, Tony Aimo, for establishing a body known as the Hunstein Range Resource Foundation which aimed to acquire one or both of the buildings in Ambunti station which WWF had bequeathed to the provincial government. Aimo was quoted as saying that:

> I used to think that the forest conservation programme was purely conservation, but that is not true … it is about sustainable development of the forest resource and protection of fragile biological resources, and the empowerment of local resource owners with World Bank funding support … I now call on the National Government and Treasury Minister to immediately conclude all agreements with the World Bank on the forest conservation program so my people can move on with the development of the April-Salumei FMA. (Anon. 2004)

The 'World Bank funding support' to which Aimo made reference would have been money earmarked by the Global Environment Facility for the Mama Graun Conservation Trust Fund as part of a 'forestry and conservation project' which was still being negotiated with the PNG government and which never actually eventuated. The trust's chairman responded to this piece of news by saying that he and his board had yet to seek or receive project proposals from anywhere in PNG, and had certainly not committed any funds to an April Salumei project (Power 2004).

The Springs of Awareness

The earliest document I have sighted in which the April Salumei area is clearly designated as the site of a possible REDD project is a letter from Theo Yasause to Jackson Yagi that was written in September 2008. In this letter Yasause said:

> I am in receipt of your landowner resolutions empowering both you and me to act on behalf of the Landowners to produce and sell carbon credits in the Forest Management Area … I hereby accept the responsibility as Designated National Authority (DNA) and will assist you with identifying an investor, in particular Earth Sky Ltd, a Company who is partnering the DNA Office to assist with this funding package.

In November 2008, Yasause issued a statement in which he said that 'Papua New Guinea recognizes and supports the April Salome Project as a Voluntary Emissions Reduction project (VER) and confirms that it meets the Sustainable

Development Objectives of Papua New Guinea'. Later that month, it seems that Yasause received a letter from Prime Minister Michael Somare, in which the latter said:

> I make reference to series of meetings with Hustein [sic] Range Landowners and Chairman and Director of April Salome Forest Management Area (FMA) project who have been waiting for payments for eco-system services since the initial discussion in 1996 … The landowners have since pursued two parallel processes to either sell their carbon credits or to undertake logging themselves which they are legally entitled to do which would deforest the area for sale of logs. They have this legal entitlement since 1996 but have voluntarily not acted despite entering into a legal case in 1998 and subsequently winning the case in 2007.

Copies of all three documents were included in the same project design document that featured the earlier memorandum to the prime minister on the subject of 'Collaboration with Earth Sky and Climate Assist PNG' (Earthsky 2008). The aim of the whole exercise was to show how the ASSFMP could be recognised as a genuine REDD project by the custodians of two international standards known as the Voluntary Carbon Standard and the Community, Climate and Biodiversity Standard.

It is hard to tell when Jackson Yagi first started to discuss the ASSFMP with Theo Yasause and representatives of Earthsky Ltd and South Pole Carbon Asset Management Ltd. In December 2008, the ASSFMP proposal claimed that scientists at the University of Papua New Guinea (UPNG) were already developing an estimate of carbon stocks in the April Salumei area (Earthsky 2008: 7). In April 2009, a stakeholder consultation process was initiated at a meeting between Yasause, Yagi, and one of UPNG's senior professors, Chalapan Kaluwin (Chem-Clean et al.: 2009: 43). The consultation process hinged on a series of four meetings held in May that year — the first at the aptly named Sepik Center of Hope in Wewak, the capital of East Sepik Province, and the rest in three of the villages in the April Salumei area. Each meeting seems to have lasted around three hours and involved an exchange of views between local stakeholders and representatives of a national company called Chem-Clean Ltd. UPNG staff and representatives of South Pole were party to the process of organising and recording these exchanges. Chem-Clean submitted draft minutes of the four meetings to the OCC within a fortnight of conducting them (Chem-Clean 2009), but the South Pole people were clearly dissatisfied with the way in which responses to local people's questions had been represented in this document. As a result, they compiled a second report which included a running commentary on the validity of responses offered by the Chem-Clean consultants and a record of subsequent written exchanges with other stakeholders, including WWF personnel (Chem-Clean et al. 2009). This second report seems to have been completed around the end of July 2009.

The likelihood of distorted communication at the four consultation meetings may be inferred from the text of the public notice which advertised their subject matter. For example, the English-language text of the notice which appeared in a national newspaper said that '[l]eakage effects will be addressed by supporting sustainable plantations and an[y] remaining leakage will be addressed by an econometric analysis of supply elasticity, leading to a conservative adjustment of emission reductions' — a statement which understandably failed the translation test in the Tok Pisin version on the bottom half of the page (OCCCT 2009). Chem-Clean's own minutes record some of the distortions that actually took place. For example, at the meeting in Wewak:

> One audience member was asking 'What are other risks or bad things from Carbon trade after talking about good things?' This shows that people are still ill informed about issues on climate change and carbon trade. Anyway, we manage to explain to the crowd the negatives or risks of carbon trade such as leakages, additionality, permanence, contractual obligations, slicing of credits because of deforestation, socio-economic impacts, fluctuations of carbon prices in the stock markets etc. (Chem-Clean 2009: 6)

A question posed in one of the village meetings elicited an equally puzzling response:

> 'What are the penalties for non-compliance to legal agreement?' We responded that the buyer can slice our revenue if there is deforestation. They have to be responsible for their act and do the right things to protect their environment and forests. They can educate their children and make proper land-use plans or even ask the developer to included [sic] buffers that can offset any leakages, additionalties [sic] and permanence. (ibid.: 16)

Given that most of the village conversations were conducted in Tok Pisin, it is hard to imagine what was actually said here, let alone what was understood. It is doubtful whether any of the local villagers would have ended up with a better understanding of the relationship between land-use plans and the technicalities of carbon sequestration in light of the Tok Pisin version of the following conversation in another village meeting:

> A landowner was concerned that they have a WMA within the FMA area of 520,000ha. The 126,000 currently and its carbon credits of 2.2 million GtCO2e is too small. He was concern [sic] that if the 126,000ha does include both the WMA and FMA. We elaborated that under REDD, the 200,000ha of WMA is not qualified because it was already 'protected'. This means only the FMA area of 520,000ha is eligible. However, we mentioned that since the WMA is inside the FMA area and there are still legal issues between PNG Forest Authority and the Department of Environment and [C]onservation, REDD may be the only solutions to both problems and could be easily resolved. Hence, the landowner urged the government to consider the 520,000 ha a REDD project to avoid confusion. (ibid.: 8–9)

Even the second report on the consultation process failed to clarify the physical extent of the area from which carbon credits would be extracted.

The additions made by the South Pole people to the second report were less concerned with the quantification of leakage, additionality and permanence than with the question of how REDD project revenues would be distributed. When someone 'in the crowd' at the Wewak meeting asked about 'benefit sharing mechanisms', the Chem-Clean consultants apparently said that they had 'no idea', and referred the question to the HRH representative, who 'stated that they will get over 60% whilst the rest will go to the government and the developer' (Chem-Clean 2009: 7). It is not clear from the draft minutes whether 'they' would be the landowners or the HRH directors, so members of 'the crowd' might have been forgiven for concluding that the landowners would get nothing at all. As for the 'developer', the consultants told the people present at one of the village meetings that '2.2GtCO2e was already sold to Earthsky' (ibid.: 9). This prompted a 'landowner company representative' to allege that 'some money valued at USD25 million was already misused by the government when Earthsky [was] already paid upfront', and then go on to say that bank statements were now being collected 'to justify and validate their claim' (ibid.: 10).

At this juncture, the South Pole people added a note to the second report to say that they did not think Earthsky had so far made any payments to the PNG government. This note immediately went on to say that all of the revenues from the sale of carbon credits would accrue to a body called the April Salome Foundation (ASF), which would spend 80 per cent of the money on projects in the FMA area, 18 per cent on other projects, and would contribute the remaining 2 per cent to the UN Central Emergency Response Fund (Chem-Clean et al. 2009: 20). The second report alluded to an 'organisational chart' which represents this form of benefit distribution (ibid.: 16), and although this chart is not attached to the report itself, it was reproduced in a thesis written by Matthew Leggett, who conducted fieldwork in the April Salumei area between May and July 2009 under the auspices of WWF (Leggett 2009: 52). The chart is at variance with the note, for it shows that gross revenues from the sale of carbon credits would accrue to Earthsky, which would then subtract an unspecified amount to cover its own costs and the costs associated with provision of 'technical support' to HRH. The ASF would then receive 80 per cent (not 100 per cent) of the net revenues, and would spend this money on projects proposed by the ILGs holding shares in HRH, while 18 per cent would be paid to the PNG government (not the ASF) as a sort of tax to cover the cost of 'projects' outside the April Salumei area. Leggett's investigation of the ASF website (which no longer exists) raised a suspicion that HRH directors might be involved in the management of that body, 'which if correct would represent a clear conflict of interest' (ibid.: 53).

In light of all these conjectures, it is hard not to sympathise with a woman who was said to have become 'emotional' at one of the Chem-Clean village meetings, wanting the government to 'realise that many men in Port Moresby cheat us', while insisting that the villagers themselves 'have all the resources and we are not stupid'. The consultants took this to mean that 'the villagers are happy in their [villages] though they lack basic government services and infrastructures' (Chem-Clean 2009: 13).

Happy or otherwise, there was evidence in Chem-Clean's draft minutes that local villagers had some rather odd ideas about the carbon trade. One villager was reported to have said that 'things such as sicknesses, deaths of trees caused by bad gases, remove [sic] of good air from their forests have created panic and false alarm among people' (Chem-Clean 2009: 9), while another reportedly said that 'we thought a tower will be build [sic] to capture all bad gases and these gases will be packed and sold for cash' (ibid.: 14). Leggett's subsequent research in the area found that Chem-Clean's consultants had failed to dispel such ideas, and might well have encouraged their diffusion. Local knowledge was now encapsulated in the idea that transactions in forest carbon involved the 'work of selling air' (*wok bilong selim win*) or else the collection of 'sky money' (*mani bilong skai*) (Leggett 2009: 56). Two of his informants asked questions in Tok Pisin which Leggett translated as follows:[4]

> What kind of pipes or other things do we need to collect the carbon? Will this company South Pole come here and build a factory to collect the carbon? Or do we need to collect it in plastic bags and go to Wewak or what? We don't understand. It sounds like hard work …

> Where does this wind (carbon) come from? It comes from the ground and the forest belonging to us. If South Pole come here and destroy our forest and our resources and steal our air I'm really worried. If they eat up our air, our children are going to die — what will they breathe? (Leggett 2009: 57).

The second report on the consultation process took due note of Leggett's findings, and concluded that 'key challenges such as the legitimacy of representation by the landowner company, details on benefits and equity sharing, transparency, accountability, general information on carbon markets and current and future land use issues will have to be addressed to ensure a proper sustainable implementation of the project' (Chem-Clean et al. 2009: 37).

---

4   These two quotations are reproduced in Leggett and Lovell (2012: 125).

Survival of the Fittest

In the wake of all the scandal about carbon cowboys and the carbon cargo cult, there was a notable decline in the number of foreign companies willing to invest any more time or money in the development and promotion of REDD projects in PNG. Earthsky, South Pole and Carbon Planet were three of the companies that vanished from the national scene at this juncture. Nevertheless, a company called Rainforest Project Management Ltd (RPM) took on the role of developing the April Salumei Sustainable Forest Management Project — with April's name correctly spelt.[5] The ASSFMP proposal was submitted to the Climate, Community and Biodiversity Alliance (CCBA) in June 2010, and its validation was approved one year later.

The 2010 version of the ASSFMP proposal had no individual authors, but listed a number of 'project development partners', including HRH, the Prime Minister's Department, the OCC, the PNG Forest Research Institute (a branch of the PNGFA), and the UPNG Centre for Climate Change and Sustainable Development (RPM 2010: 7). The document was not only posted to the CCBA website, but also to the website of a body called the Rainforest Management Alliance, a body that seems to have been created by the ASSFMP's 'project manager', Stephen Hooper, who also represents the 'project implementer', RPM. The Rainforest Management Alliance website featured four other proposals for carbon trading projects in PNG (and none for projects in any other country), but there is no evidence that any of the other four proposals were ever submitted to the CCBA. The space occupied by the ASSFMP proposal also came to feature some additional documentation: a letter of support apparently written by the prime minister's legal adviser to the local MP, Tony Aimo, in July 2010; another one seemingly written by the government's acting chief secretary to Stephen Hooper himself in August 2010; two undated press releases in favour of the project from Tony Aimo and the forests minister, Belden Namah; and a copy of the Ambunti-Dreikikir District Development Plan (in which the project was not mentioned).[6]

The 2010 version of the project proposal claimed that a new round of meetings had been held with local community members and the chairmen of local land groups and landowner companies between October 2009 and January 2010, as a result of which 163 land group chairmen had supposedly signed consent forms (RPM 2010: 96). The local landowners were now said to be divided between

5   Earthsky's disappearance from April Salumei was more apparent than real, because RPM turned out to have the same registered address in the British Virgin Islands.
6   The Ambunti-Dreikikir district or electorate is the one that has been represented in parliament by Tony Aimo between 2002 and 2013. It includes the Ambunti and Tunap-Hunstein LLG areas, and thus includes about 80 per cent of the people with customary rights in the April Salumei forest area.

'four main cultural groups', each with a specified number of ILGs holding shares in a different landowner company, and these four companies were said to be shareholders in HRH (ibid.: 24). The land groups whose chairmen refused to sign the original FMA were now said to be among the 44 land groups that had agreed to let one of the four landowner companies (Niksek Samsai Resources Ltd) act on their behalf. Information about the social, economic and political organisation of the area seems to have been derived from a report written by a company called Tack Realty Ltd in 2004, but no proper reference is given to this report, nor have I sighted a copy of it. The proposal made provision for another three or four years' work to 'update' land groups and map their boundaries, and 'once it is done it will provide what will be possibly the most robust database and understanding of traditional landownership in PNG' (ibid.: 89).

Tack Realty's earlier investigation of this subject seems to have been commissioned by HRH as evidence in support of its compensation claims. The ASSFMP proposal cited this report as evidence that the opportunity cost to local landowners of keeping loggers out of the FMA area was more than US$188 million, and then said that if the landowners were to succeed in their ongoing efforts to secure K58 million in compensation from the national government, this would not entail any relinquishment of their 'logging rights' or their 'right over the biodiversity' of the area (RPM 2010: 39). The proposal used PNGFA figures to determine that the 'gross forest area' covered by the FMA was around 177,000 hectares, and if a 15 per cent 'buffer zone' were to be subtracted from this figure in accordance with the National Forestry Development Guidelines, that would leave a 'net production area' of around 150,000 hectares that would have been logged under a timber permit (ibid.: 63). Nevertheless, it is the larger of these two areas that was described as the 'carbon accounting area' (ibid.: 65), from which all agricultural activity, including the practice of shifting cultivation, would be banned under the terms of the project proposal, so people owning land within this area would henceforth have to make their gardens somewhere else (ibid.: 93). There was a glimmer of recognition that this could cause a problem for the distribution of landowner benefits from the project, because 'the population distribution map shows the population is not evenly distributed and the carbon accounting area to be least populated', but this problem was thought to be solved because 'we have assumed an even distribution to be conservative' (ibid.: 65) — whatever that meant.

By the end of July 2010, the CCBA website contained six comments on this version of the project proposal. These included comments from Theo Yasause (in his private capacity), Wari Iamo (in his official capacity), Matthew Leggett (on behalf of WWF), and Martin Golman, who was the deputy director of the Forest Research Institute. All comments except the one made by Yasause were critical of the proposal. Leggett's comment included a copy of a published press release in which Iamo denied that the OCC was a 'project partner' or was ever even approached for its 'endorsement, support or partnership' (OCCD 2010). Iamo's own comment consisted of a letter reiterating a recent decision of the National Executive Council 'that voluntary trading of forest carbon is inadvisable and premature under present arrangements', and attached legal advice to the effect that PNG's laws did not yet allow for the creation of negotiable property rights in forest carbon (CCBA 2010: 31–8). Golman's comment was the longest and most critical of all, possibly because he was, as he said, partly 'representing himself as a prominent landowner of Malu village in the project vicinity and an expert in sustainable forest management practices and carbon assessment in April Salumei',[7] but also because he had some responsibility for negotiating the FMA with local landowners in 1996 in his capacity as an official of the National Forest Service (ibid.: 9–10). Golman maintained that many of the landowners in the April Salumei area had never consented to being represented by Jackson Yagi and HRH in the negotiation of any project at all, nor had they ever received any share of whatever money HRH had managed to secure by way of 'compensation' from the state for the opportunity costs of forest conservation (ibid.: 13). He went on to propose that any further research relating to forest management in the area of overlap between the FMA and the WMA should be conducted by the Samsai Niksek Forest and Carbon Foundation, founded by himself, in partnership with the Forest Research Institute (ibid.: 18).

The Crowning Moment

Despite this negative commentary, April Salumei was still crowned as the only pilot REDD project in an area covered by an FMA. The 'validation' of the project was undertaken by a Californian body called Scientific Certification Systems, which contracted a New Zealand company called Environmental Accounting Services Ltd to check the project's compliance with the CCBA's 14 project design standards. The validators did most of their work between June and August 2010. This included an evaluation of the negative comments posted to the CCBA website and the conduct of 'community meetings' at three locations in the FMA area (SCS 2011: 6). Their report contains 41 instances of 'non-conformance' that required some modification of the project design document, but many of these were minor and easily dealt with by the project proponents. Once the project document had been revised to the satisfaction of the auditors, a 'validated copy' was released for public consumption. This time the authors were shown to be Stephen Hooper and a group of seven Papua New Guineans, all of whom had at one time or another been employed by UPNG, although one, Simon Saulei, had

---

7   He had recently completed a PhD thesis on this topic at The Australian National University.

returned to his former post as director of the Forest Research Institute at the beginning of 2009. Both documents, along with a 'monitoring plan' requested by the validators, were posted to the CCBA website.[8] They also appear on the website of another body called the Pacific Forest Alliance, which is also one of Stephen Hooper's creations.[9]

The most significant modifications of the project design to emerge from the validation process were concerned with problems of authorisation and ownership. The first of these problems was to determine which local people had some entitlement to be consulted about the project, or be made aware of it, or give their consent to it, or share in whatever costs and benefits it might generate. The 'carbon accounting area' was now confined to the 'net production area' (around 150,000 hectares) that would be logged if a timber permit were granted over the FMA area (around 520,000 hectares), but the REDD 'project zone' was now expanded beyond the boundary of the FMA area, not only to include a 10-kilometre buffer around its northern, eastern and southern flanks, but to the banks of the Sepik River in the north and to the East Sepik provincial boundary in the south (Hooper et al. 2011: 33).[10] This meant that it included the whole of the WMA that lay outside the FMA area (Figure 8.1) and probably covered about 900,000 hectares altogether. In the previous design document, the 10-kilometre buffer zone was thought to contain people who might be impacted by the project (RPM 2010: 26), but the validators thought that the 'Project zone boundary should ensure that it encompasses the boundaries of the four land holder companies', and should eliminate the possibility of an ILG land boundary 'extending past the Project boundary' (SCS 2011: 57). In an effort to comply with this peculiar requirement, the proponents now said they would give due consideration to 'traditional boundaries that extend outside the project area and into the project zone', and were apparently prepared to consider the residents of the larger project zone as people who might have use rights, if not ownership rights, within the FMA area (Hooper et al. 2011: 30–1). No attempt was made to count the number of extra people resident in this larger area,[11] or to demonstrate that they had been made aware of their new situation.

The revised project document did not provide any additional information about the number of people who had traditional rights to make use of the carbon accounting area in ways that might detract from the amount of carbon contained within it, but the former threat of prohibition was replaced with

---

8   www.climate-standards.org/category/projects/oceania/ (viewed April 2015).

9   www.pacificforestalliance.org/ (viewed April 2015).

10   For some unknown reason, no buffer zone was created on the western banks of the Leonard Schultze River, which constitutes the western boundary of the FMA area.

11   The project proponents have consistently represented the resident population of the FMA area alone as being 7696 in 2000 (RPM 2010: 33; Hooper et al. 2011: 39). In the absence of a detailed breakdown of this figure, I am unable to explain the discrepancy between this number and my own calculation (Table 8.1).

a risk assessment and mitigation matrix (Hooper et al. 2011: 100–9). The 41 risks evaluated in this section of the document included such things as 'population growth', 'agricultural pressure', 'hunting of endangered or threatened species', 'small scale clearing of land', 'regular burning of grasslands' and 'access to areas fundamental for meeting basic needs of the local community'. But apart from the possible impact of hunting on biodiversity, none of these was considered as problematic as risks like 'ILG leadership struggle', 'lack of knowledge and skills in business administration', 'inappropriate use of funds', 'financial benefits not shared equally', or 'slow delivery of the project'. Most of the risks identified in the matrix would be offset by a mixture of 'awareness and monitoring' activities on the part of local 'community stewards' (ibid.). But the auditors thought that these activities would have to be framed by 'an appropriate methodology for the identification of impacts on the community' (SCS 2011: 66), as well as a 'community monitoring plan' (ibid.: 68). The revised project document therefore includes another new section on social impact indicators that is remarkable mainly for its observation that the biggest impact of the project on the local people is likely to be 'the identification of the true landowners and the subsequent distribution of benefits to them' (ibid.: 158).

In their own process of consultation with local communities, the validators discovered a popular belief that an unequal distribution of benefits to ILGs resulting from the unequal size of their landholdings would probably lead to conflicts that might (or might not) be resolved in many different ways (SCS 2011: C-20). One might imagine further scope for conflict if the distribution of benefits did not match the distribution of customary rights to exploit the natural resources in the carbon accounting area. But Stephen Hooper shied away from the idea that benefits should be matched to impacts or opportunity costs:

> It has also been agreed by the landowners and documented in the PDD [that] all funds emanating from the Project will be distributed equally amongst all landowners. That is every man and woman over 16 years of age will share equally. This we believe is a fair system. Payments will be made directly to the Incorporated Land Groups and not through Hunstein Range Holdings. (SCS 2011: C-7)

The reference to HRH reflects the well-documented local 'awareness' of its previous failure to distribute benefits in an equitable manner (SCS 2011: 11). The statement as a whole implies that this failure would henceforth be compensated by payments of cash to each one of the 163 ILGs whose chairmen consented to the project at the end of 2009, the amounts of which would vary in proportion to the number of their adult male and female members. But aside from a passing reference to compensation payments 'in cash and kind' in the project design document (Hooper et al. 2011: 156), the general presumption was that direct benefits to local communities would take the form of employment with the REDD project itself and a set of specific 'social and infrastructure projects' approved by the board of the ASF (ibid.: 118). There is no way the REDD project would produce 163 jobs. A much smaller number of local workers would apparently be employed by one or other of the four new landowner companies (ibid.: 113), and their executives would only achieve a voice on the ASF board through their membership of the HRH board (SCS 2011: B-5).

The validation process failed to clarify the nature of the property rights vested in HRH and the consequent rights of its directors to participate in the REDD project. The validators decided that the project was initially authorised by an agreement between HRH and Earthsky that dated back to May 2009 (SCS 2011: 22), and Hooper then agreed to describe Earthsky as the 'project originator', which had appointed RPM to develop and manage the project on its behalf (ibid.: 63). At the same time, the validators said that they agreed with RPM that HRH was 'the current legal entity that has been recognised through the PNGFA and the PNG Supreme Court as having the resource rights to the FMA and are therefore a significant Project stakeholder' (ibid.: B-5). However, there is no evidence that the PNG Supreme Court or any other court had vested these rights in HRH, and any decision along these lines would be totally inconsistent with the Forestry Act, which clearly states that an FMA is an agreement by ILGs — not a landowner company — to transfer timber harvesting rights to the PNGFA.

Jackson Yagi's claim for compensation had been based on the failure of the PNGFA to convert the FMA area into a logging concession from which he and other landowners might derive some material benefit. Hooper himself told the validators that the REDD project was conceived as part of an 'out of court settlement' of this legal dispute (SCS 2011: 72). What was at stake here was not the ownership of the timber resources (or the forest carbon) in the FMA area, but the degree to which the declaration of the Hunstein Range WMA justified the PNGFA's failure to grant a timber permit to a logging company. The circumstances of the settlement are unclear,[12] but proponents of the REDD project needed to prove that the WMA had been invalidated in order to acquire the magical substance known to the world of carbon trading as 'additionality'. In other words, they had to make a case that the FMA, or more specifically the carbon accounting area, *would have been logged* in the absence of their project (ibid.: 63–5). If the National Forest Board was to take the existence of the WMA as evidence that some local landowners had never consented to the FMA, and use this as a pretext for not granting a timber permit, there would be

---

12   According to Leggett (2009: 46), it was the Department of Environment and Conservation that settled the HRH claim in 2007, but a commission of inquiry into the Department of Finance determined that HRH received the first payment in respect of its claim in 2005 (GPNG 2009: 4041).

no additionality.[13] Since that was the board's position before Jackson Yagi made his claim against the state, it was essential for the project proponents to award him a victory of some sort (ibid.: 14). But how could that be reconciled with the need to secure the support of all the local landowners? Stephen Hooper had two solutions to this problem. One was to wash his hands of it by refusing to become involved in 'local level politics' and insisting that 'the structure and make up of representative companies must be determined by the people living in the villages on a daily basis' (ibid.: C-4). The other was to make an arrangement for the PNG Electoral Commission to supervise the democratic election of the boards of each of the four — which now became five — new landowner companies within six months of project validation, and to require that each of their boards have two representatives on the board of HRH.[14] Jackson Yagi told the validators that he was happy with this arrangement (ibid.: 58).

If these measures promised to deliver a new sense of joint ownership among the local landowners, other measures were taken to enlarge the circle of support from national stakeholders. Although WWF representatives declined an invitation to be involved in a further round of community consultation in August 2009, one of the organisation's national employees later took leave from his job to identify gaps in the state of landowner awareness (SCS 2011: C-3). A local NGO, Partners with Melanesians, was engaged to be the project's dispute resolution authority (ibid.: 24), and both organisations were invited to join an independent review body to advise the ASF's board of trustees (ibid.: C-7). Martin Golman's critique, which the validators ascribed to a 'degree of professional jealousy' (ibid.: B-3), was quickly muted when the forests minister, Belden Namah, expressed his own support for the project in August 2010 (ibid.: 64). Although Namah then crossed the floor of parliament to become leader of the opposition, his erstwhile support was sufficient for work on the April Salumei project to be funded in the PNGFA's budget for 2011.

This meant that Wari Iamo and the OCC began to look somewhat isolated in their public denial of support for the project. The validators sought to neutralise this source of opposition by suggesting that Iamo may not have been the executive director of the OCC when he published his adverse advertorial in July 2010 (SCS 2011: 33), while Stephen Hooper argued that the project did not need support from the OCC anyway, so long as it had the backing of the prime minister, the forests minister and the local MP (ibid.: 63). Iamo was certainly the secretary of the Department of Environment and Conservation (DEC), and

---

13   At one point in their report, the validators seem to have thought that the WMA was still officially recognised but had somehow been excluded from the carbon accounting area (SCS 2011: 73), but there is nothing in the REDD project design document to support this notion.

14   A fifth landowner company was established in September 2010 because of a split between two sets of land groups in the WMA that had formerly been shareholders in Niksek Samsai Resources Ltd (SCS 2011: C-5).

there was no doubt that he had been appointed as the acting executive director of the OCC when Theo Yasause was suspended, which meant that the OCC had looked like a branch of the DEC since July 2009. If he no longer occupied this extra position in July 2010, then it seems rather strange that he chaired the multi-stakeholder workshop convened in Port Moresby's Hideaway Hotel in February 2011 to endorse the National Programme Document (NPD) (that would release US$6.4 million from the UN Collaborative Programme to implement the first, three-year phase of PNG's 'REDD+ readiness roadmap' (UN-REDD 2011). The managing director of the PNGFA informed this meeting of his agency's intention to implement the April Salumei project as a pilot project, but he did not chair the meeting, nor was the project mentioned in the NPD.

This was just one moment in an ongoing struggle between the PNGFA and the OCC (or DEC) for control of the forest carbon policy process. Local environmental NGOs made an unaccustomed alliance with national forestry officials in this struggle because they saw the reformed version of the OCC as a puppet of Kevin Conrad and the group of highly paid foreign consultants whom he had introduced to this process immediately after the Copenhagen climate change conference at the end of 2009 (Lang 2010b; Babon 2011; Filer and Wood 2012; Babon et al. 2014). Supporters of the April Salumei project were reluctant to endorse the NPD while there was no prospect of an end to this form of foreign interference. Their wish was granted sooner than they might have expected. In August 2011, Michael Somare was ousted as prime minister. His replacement, Peter O'Neill, appointed Belden Namah as his deputy, restored him to his former portfolio, and put him in charge of the OCC as well as the PNGFA. Shortly after his appointment, Namah announced the dismissal of Kevin Conrad from his position as PNG's climate change ambassador on the grounds that he had little or no knowledge of the 'culture, tradition and lifestyle of the people', and therefore could not deal with 'landowner issues' (Waima 2011). Meanwhile, the foreign consultants had already disappeared because their contracts had expired or their fees had not been paid. Wari Iamo retained his position as acting head of the OCC for the time being,[15] but ceased his pursuit of the case against April Salumei.

In the months that followed, the project monitoring plan was finalised (PFA 2012), a new study was undertaken to assess the feasibility of attaining compliance with the Voluntary Carbon Standard (Green 2012), and some additional work was done on the measurement of local carbon stocks. For some unknown reason, the Voluntary Carbon Standard Feasibility Assessment added 100,000 hectares to the size of the net forest production area (Green 2012: 5), but it did not

---

15   Like many other public servants, he resigned from office in March 2012 in order to contest the forthcoming national elections. He was not elected.

provide any new information about the customary landowners. One interesting feature of this report was the inclusion of a new organisation chart in which the chairmen of the five landowner companies would be members of a new body called the April Salumei Working Group, whose interaction with project management would not be mediated by the board of HRH (ibid.: 8). It was also stated that local landowners would 'receive a minimum of 60% of all sale proceeds directly', and that this 'paradigm shift to landowner involvement and payments' had been authorised by the national government. While there was no evidence of any sales proceeding, two motor boats and K200,000 had apparently been delivered to the working group to 'kick start local community activities' (ibid.: 9).

All this happened before the national elections in the middle of 2012. Thereafter, Peter O'Neill retained his position as prime minister, Tony Aimo retained his position as the local MP, while Belden Namah returned to his former role as leader of the opposition. The new minister of forests and climate change was another former forests minister, Patrick Pruaitch. Pruaitch is an ally of Michael Somare, who now became the governor of East Sepik Province and declared his support for the ruling coalition. The National Executive Council reiterated its own endorsement of the ASSFMP in December 2012. To celebrate this moment, Tony Aimo appeared on national television with the chairman of the April Salumei Working Group, the local project manager, and a king-sized replica of a cheque for K300,000 to be divided equally between the five landowner companies. At the same time, Aimo told a national newspaper that April Salumei would be 'the biggest carbon trade project in the world to date' (Anon. 2012b).

Conclusion

April Salumei is not in fact the only forest carbon project that was still alive in PNG in 2013. The PNGFA claimed to have four other 'REDD+ pilot projects' in the 2012 draft of its National Forest Plan (GPNG 2012a: 18), while five NGOs were said to have received grants from the Australian government to develop 'REDD+ demonstration activity concepts' (Babon 2011: 5). However, April Salumei was the only survivor of the many projects that were being promoted back in 2008, and the only one that seemed to have any prospect of producing marketable carbon credits instead of simply being funded by the national budget or foreign aid. The other projects in the National Forest Plan all involved afforestation or reforestation or reduced impact logging, there was no detail on the activities being undertaken or proposed, and none of them was clearly tied to an area where the government had already acquired the timber harvesting rights without passing them on to a logging company.

April Salumei's survival in a process of natural selection may have something to do with her physical qualities. Her biodiversity values do not exceed those of any other forest area in PNG, but a good case can be made for them to exceed those of any other FMA area where logging has not yet been authorised. This was a case which WWF made to the government back in 1996, and a case which the REDD project proponents had been making more recently. The CCBA awarded the project a gold medal for 'exceptional biodiversity benefits', but not for exceptional climate or community benefits (SCS 2011: 55; Hooper et al. 2011: 171).

April Salumei's survival also had something to do with her status as a form of property and her claim to the magical power of additionality. The PNGFA was less convinced of its capacity to authorise logging in some of the other FMA areas because the original agreements with local landowners were even more questionable than they were in this case. Could it therefore be argued that the ASSFMP thrived on an exceptional level of local community support or an exceptional ability of external agents to mobilise community participation? There is not enough evidence to evaluate recent interactions between landowner representatives and outsiders like Stephen Hooper, but there is little reason to credit the effort made to produce 'landowner awareness' of the REDD project in 2009. That was surely a false start. If there had been an increase in local understanding and appreciation of the project since that time, the evidence suggests that it was far from being equally distributed among different groups of landowners. It is hard to imagine that a local 'community' that may now contain 5000 adult members (if one includes absentee landowners like Martin Golman) could possibly speak with one voice when they have 13 or 14 vernacular languages. The political, social and cultural fragmentation of the April Salumei area is a huge — and possibly insurmountable — challenge for a comprehensive demonstration of free, prior and informed consent (Anderson 2011; Leggett and Lovell 2012). The recent semblance of consensus may be due to a common but fragile sense of anticipation, which could either break down if the anticipated benefits fail to materialise or break down if they do materialise and are (thought to be) unequally or unfairly distributed.

Perhaps the level of community support has been less significant than the level of support from provincial stakeholders, including the local MP and the former prime minister. Theo Yasause also hailed from East Sepik Province, and there is no doubt that he showed greater enthusiasm for April Salumei than he did for any other REDD project proposal during his time in charge of the OCC. He also had good reason to think that Michael Somare — and therefore Kevin Conrad — would prefer to see PNG's first REDD project developed in their own province. Yasause's own preference no longer matters, but when Somare met the forest-friendly Prince of Wales on the occasion of his mother's diamond jubilee in November 2012, Somare proposed to honour the occasion by making East Sepik a REDD 'demonstration province' (Anon. 2012a). Conrad and Pruaitch were both in attendance to welcome this initiative. On the other hand, there is no evidence that April Salumei's most active supporters made an effort to get Somare's stamp of approval — let alone that of Conrad or the Prince — after Yasause was removed from office in 2009.

I think a stronger case can be made for the importance of support from a group of national stakeholders who are members of what I call the 'conservation policy community' (Filer 2005). As we have seen, the original members of this group were based at UPNG in 2008, but membership later expanded to include people employed in the Forest Research Institute, the PNG University of Technology, the Vision 2050 (long-term planning) branch of the Prime Minister's Department, and one of the local NGOs that are affiliated with the PNG Eco-Forestry Forum. Most of the individuals in this group do not hail from East Sepik Province, but one thing they had in common was their opposition to Kevin Conrad's control of the forest carbon policy process. Beyond that, they had a lukewarm relationship with the big international NGOs that had taken an interest in this process, despite (or possibly because) some of them were previously employed by these organisations. This might help to explain why WWF did not show much enthusiasm for the ASSFMP, even when foreigners like Stephen Hooper tried to gain its endorsement. As we have seen, the pursuit of 'additionality' entailed an attack on the validity of the WMA which WWF had previously tried to establish as a defence against the prospect of logging, but there was also a kind of resource nationalism at work in claims to ownership of April Salumei. In this respect, she became a stage from which her national supporters sought to fill the vacant space that Conrad should have left behind him.

It is hard to say whether the factors that explain the relative success of the April Salumei project to date will explain its continued success in future, or whether its further development will serve to clarify the relative importance of these different factors. It seems unlikely that this project alone will generate significant revenues if no further progress is made with the national 'roadmap' whose engineering is still vested in the OCC, but progress has been rather slow. In the two years after the UN Collaborative Programme agreed to spend US$6.4 million on PNG's roadmap, the OCC only managed to produce a second draft of its Climate Compatible Development Policy (GPNG 2013a), a working draft of a Readiness Preparation Proposal (GPNG 2013b), and a draft bill for a Climate Change Authority Act. It managed to finalise a set of guidelines and an institutional framework for the official approval of REDD+ projects (GPNG 2012b), including draft guidelines for securing free, prior and informed consent for such projects (GPNG 2013c), but it is not clear whether the April Salumei project has passed through this approval process or whether its endorsement by the PNGFA and the National Executive Council has itself been counted as sufficient evidence of government authorisation. The Readiness Preparation Proposal says that project activities in the area 'will be determined after a development option study' (GPNG 2013b: 60). This is a type of study undertaken under the terms of the Forestry Act, and if another one is undertaken, it will be the third or fourth to be undertaken since the timber harvesting rights were first acquired by the state in 1992. The local landowners may have to wait a while longer for their carbon credits to be realised.

Acknowledgements

Thanks to Andrea Babon, Paul Chatterton, Stephen Hooper, Matt Leggett, Dave Melick, Rune Paulsen, James Sabi, Gwen Sissiou and Mike Wood for comments and information. None of them bears any responsibility for the argument contained in the paper.

References

Anderson, P., 2011. 'Free, Prior and Informed Consent in REDD+: Principles and Approaches for Policy and Project Development.' Bangkok: RECOFTC (The Center for People and Forests).

Anon., 2004. 'MP Aimo Gets Behind April-Salumei Project.' *The National*, 30 January.

Anon., 2008. 'Carbon Trading or Carpet Baggers.' *Post-Courier*, 22 November.

Anon., 2009a. 'Aihi Urges PM to Launch Probe into Climate Change Office.' *The National*, 9 June.

Anon., 2009b. 'Yasause Slams Media Report.' *The National*, 9 June.

Anon., 2009c. 'Group Awaits K500,000.' *The National*, 9 July.

Anon., 2011. 'Big Sister's Carbon Tax, a Catalyst for Carbon Cowboys?' The Masalai Blog, 28 August.

Anon., 2012a. 'HRH Meets on Forestry.' *Post-Courier*, 13 November.

Anon., 2012b. 'Forest Owners Ready to Benefit.' *The National*, 6 December.

Babon, A., 2011. 'Snapshot of REDD+ in Papua New Guinea.' Bogor: Center for International Forestry Research (Infobrief 40).

Babon, A., D. McIntyre, G.Y. Gowae, C. Gallemore, R. Carmenta, M. Di Gregorio and M. Brockhaus, 2014. 'Advocacy Coalitions, REDD+, and Forest Governance in Papua New Guinea: How Likely Is Transformational Change?' *Ecology and Society* 19(3): 16.

Bamu, S., 2005. 'Carbon Swap Has a Bad Smell.' *Post-Courier*, 20 June.

Callick, R., 2009. 'The Rush Is On for Sky Money: The Struggle Begins to Control the Immense Wealth Certain to Flow from Carbon Trading.' *Weekend Australian*, 5 September.

Callick, R. and I. Gridneff, 2009. 'Rush Is On for "Sky Money".' *Post-Courier*, 10 September.

CCBA (Climate, Community and Biodiversity Alliance), 2010. 'April Salumei Sustainable Forest Management Project: Comments Received by the CCBA During the Validation Audit.' Arlington (VA): CCBA.

Chem-Clean Ltd, 2009. 'April-Salumei Pilot Reduce Emissions from Deforestation and Degradation (REDD) Project Initial Stakeholder Consultation Report.' Unpublished report to PNG Office of Climate Change and Environmental Sustainability.

Chem-Clean Ltd, University of PNG and South Pole Carbon Asset Management Ltd, 2009. 'April Salome Sustainable Forest Management Project: First Stakeholder Consultation Report.' Unpublished report.

Duguman, J., 2004. 'Review of the Sepik Community Land Care Project.' Unpublished report to World Wide Fund for Nature.

Earthsky Ltd, 2008. 'Project Design Document: April Salome Sustainable Forest Management.' Unpublished draft report to Voluntary Carbon Standard.

Filer, C., 2005. 'The Conservation Policy Community in Papua New Guinea.' Canberra: The Australian National University, Resource Management in Asia-Pacific Program (Working Paper 55).

Filer, C., 2012. 'Why Green Grabs Don't Work in Papua New Guinea.' *Journal of Peasant Studies* 39(2): 599–617.

Filer, C. and M. Wood, 2012. 'The Creation and Dissolution of Private Property in Forest Carbon: A Case Study from Papua New Guinea.' *Human Ecology* 40(5): 665–77.

GPNG (Government of Papua New Guinea), 2001. 'April Salumei (East Sepik Province).' Port Moresby: PNG Independent Forestry Review Team (Individual Project Review Report 25).

GPNG (Government of Papua New Guinea), 2009. 'Transcript of Proceedings, Wednesday 8 July 2009.' Port Moresby: Commission of Inquiry into the Department of Finance.

GPNG (Government of Papua New Guinea), 2012a. 'National Forest Plan' (draft). Port Moresby: PNG Forest Authority.

GPNG (Government of Papua New Guinea), 2012b. 'National REDD+ Project Guidelines.' Port Moresby: Office of Climate Change and Development.

GPNG (Government of Papua New Guinea), 2013a. 'Climate Compatible Development Policy 2013–2015.' Port Moresby: Office of Climate Change and Development.

GPNG (Government of Papua New Guinea), 2013b. 'Readiness Preparation Proposal (R-PP) – Version 11: Final Working Draft.' Port Moresby: Office of Climate Change and Development.

GPNG (Government of Papua New Guinea), 2013c. 'Revised Draft Guidelines on Free, Prior and Informed Consent for REDD+ in Papua New Guinea.' Port Moresby: Office of Climate Change and Development.

Green, C., 2012. 'April Salumei Sustainable Forest Management Project: VCS Feasibility Assessment.' Report to Pacific Forest Alliance Ltd.

Gridneff, I., 2009a. ' "Sky Money" Carbon Scheme a Hoax.' *The National*, 15 June.

Gridneff, I., 2009b. 'Aust Company Caught in PNG Carbon Fiasco.' *Post-Courier*, 31 July.

Guddemi, P., 1997. 'Continuities, Contexts, Complexities, and Transformations: Local Land Concepts of a Sepik People Affected by Mining Exploration.' *Anthropological Forum* 7(4): 629–648.

Hooper, S., C. Kaluwin, J. Duguman, G. Gowae, A. Asmann, E. Kwa, O. Gideon and S. Saulei, 2011. 'Project Design Document: April Salumei, East Sepik, Papua New Guinea.' Arlington (VA): Climate, Community and Biodiversity Alliance.

Howes, S., 2009. 'Cheap But Not Easy: The Reduction of Greenhouse Gas Emissions from Deforestation and Forest Degradation in Papua New Guinea.' *Pacific Economic Bulletin* 24(1), 130–43.

Lang, C., 2009a. 'Anatomy of a Deal: The April Salome REDD Project in Papua New Guinea.' *REDD-Monitor*, 7 June.

Lang, C., 2009b. 'Kevin Conrad on REDD, Irregularities and Carbon Cowboys in PNG.' *REDD-Monitor*, 9 July.

Lang, C., 2010a. 'Forests, Carbon Markets and Hot Air: Why the Carbon Stored in Forests Should Not Be Traded.' *REDD-Monitor*, 11 January.

Lang, C., 2010b. 'McKinsey's REDD Plans in Papua New Guinea: Nice Work If You Can Get It.' *REDD-Monitor*, 7 October.

Leggett, M., 2009. 'Selim Win Bilong Yumi': The Social Dimensions of Carbon Forestry in Papua New Guinea. Edinburgh: University of Edinburgh (MA thesis).

Leggett, M. and H. Lovell, 2012. 'Community Perceptions of REDD+: A Case Study from Papua New Guinea.' *Climate Policy* 12(1): 115–34.

Melick, D., 2008. 'On-the-Ground Role of NGOs with Emerging Carbon Market Issues: Is the REDD Emperor Actually Wearing Any Clothes?' Paper presented to seminar on 'Challenges to the National Implementation of Activities to Reduce Emissions from Deforestation and Forest Degradation (REDD)', The Australian National University, Canberra, 18 June.

Melick, D., 2010. 'Credibility of REDD and Experiences from Papua New Guinea.' *Conservation Biology* 24(2): 359–61.

Morauta, M., 2009. 'Carbon Trading and Office of Climate Change and Environment Sustainability.' *The National*, 20 July.

OCCCT (Office of Climate Change and Carbon Trade), 2009. 'Call for Stakeholder Comments and Invitation for the Meeting.' *Post-Courier*, 30 April.

OCCD (Office of Climate Change and Development), 2010. 'Voluntary Carbon Projects.' *Post-Courier*, 14 July.

OCCES (Office of Climate Change and Environmental Sustainability), 2009. 'Voluntary Carbon Agreements (VCA) Are Not Currently Supported by the Government.' *Post-Courier*, 2 September.

Pearse, R., 2012. 'Mapping REDD in the Asia-Pacific: Governance, Marketisation and Contention.' *Ephemera* 12(1/2): 181–205.

PFA (Pacific Forest Alliance Ltd), 2012. 'April Salumei Rainforest Preservation Project: April Salumei CCB Monitoring Plan.' Arlington (VA): Climate, Community and Biodiversity Alliance.

Power, T., 2004. 'Timber Project Facts.' *The National*, 9 February.

RPM (Rainforest Project Management Ltd), 2010. 'Project Design Document: April Salumei, East Sepik, Papua New Guinea.' Unpublished draft report to Climate, Community & Biodiversity Alliance.

SCS (Scientific Certification Systems), 2011. 'Final CCBA Project Validation Report: April Salumei, East Sepik, Papua New Guinea.' Arlington (VA): Climate, Community and Biodiversity Alliance.

Somare, M., 2009. 'Carbon Trading and Office of Climate Change and Environmental Sustainability.' *Post-Courier*, 3 August.

Talu, P., 2009. 'PM Under Fire Over K547mil Payment.' *The National*, 4 September.

UN-REDD (United Nations Collaborative Programme on Reducing Emissions from Deforestation and Forest Degradation in Developing Countries), 2011. 'National Programme Document: Papua New Guinea.' Port Moresby: UN-REDD.

Waima, K., 2011. 'Climate Change Ambassador in US to Be Replaced.' *Post-Courier*, 18 August.

Wynn, G. and S. Creagh, 2009. 'Forest Carbon Market Already Shows Cracks.' *Post-Courier*, 9 June.

9. Representational Excess in Recent Attempts to Acquire Forest Carbon in the Kamula Doso Area, Western Province, Papua New Guinea

MICHAEL WOOD

The rainforests of Papua New Guinea (PNG) have long been regarded as containers of still to be revealed mysteries and wonders. One persistent theme in representations of these rainforests is the lack of visibility that is found inside the rainforest. Another highlights the need for some unique kind of vision, typically provided by science, that would reveal its underlying logic and structure. A third theme is that what the forest ultimately reveals contributes to its own conservation and reproduction.

Charles Lane Poole, one of the region's earliest and most influential professional foresters, noted that describing the type of forests found in PNG was:

> ... not easy. The large buttresses that certain trees have at the base of the boles, the hanging lianas, the aerial gardens that cover the upper branches of the tall trees, all tend to distract and make a comprehensive view or description difficult. Add to these the thick stand of young trees which makes it hard to see stems of the trees more than half a chain away, and some idea of my difficulty in giving a general view of this type of forest may be obtained. (Lane Poole 1925: 6)

However, once this limited visibility was supplemented by scientific vision that moved the observer away from aesthetics, 'beautiful excrescence' and other forms of excess, and into the realm of forestry expertise, a certain forest structure could be observed. According to Lane Poole, treating the forest as a resource — timber — allowed for a specific clarity of vision:

> The complexity of the forest is much reduced when it comes to examining it from a purely forestry stand point. The often beautiful excrescence of the scene must be overlooked while the main thing — timber — is examined ... Once this is determined on, the forest becomes a simple matter to describe, for it now consists of three storeys of trees — the top one, about 100–120 feet high, is composed of trees from 7-ft girth upwards; the second of trees 5 feet in girth 50 to 75 feet high; and the third of trees 30 feet high. (Lane Poole 1925: 6)

Despite this clarity, the diversity of species, and their spatial dispersal, prevented easy estimation of timber resources. More recently the ability to estimate how much rainforest currently exists, or will exist, in PNG after industrial logging has become a standing feature of policy debates concerning 'sustainable' forestry (Nadarajah 1993; Filer et al. 2009; Shearman et al. 2009, 2010).

Supplementing these concerns of foresters, conservationists and botanists about the definition of sustainable logging in PNG, biologists have often found the animals living in the PNG rainforest difficult to see, describe and measure. As a result, the PNG rainforest remains full of excitingly unknown animals. One of Australia's more prestigious newspapers, *The Age*, ran a story entitled 'PNG rainforests reveal 200 new species' (7 October 2010). The article celebrated the 'spectacular variety' and beauty of the animals discovered by scientists working in mountain forest. The organisers of the expeditions hoped that their findings would contribute to the conservation of this type of forest. The implication was that PNG forests could reveal even more species diversity on a similar scale and that such biodiversity was intrinsically valuable. Such diversity could apparently, by itself, create conservation effects for the rainforest.

Another, rather different, conjunction of rainforest, media representations and conservation effects has emerged from the idea that PNG forests contain vast and measurable (Bryan et al. 2010; Fox et al. 2010) amounts of marketable forest carbon. As outlined in this chapter, while forest carbon had the astounding potential to avoid the destruction of PNG's rainforests, actual attempts to create productive conjunctions of forest carbon and conservation often involved illegal activities that were highly contested and easily overturned.

The story I tell of these politics primarily involves forest originally controlled by Kamula (and Doso) speakers in the lowlands of Western Province, PNG. While aspects of the events outlined highlight illegalities in the attempts to transfer property to new actors, and corporate structures that apparently represent the

Kamula and other communities, what I highlight is the way the media became a vehicle for political struggle over the development of marketable forest carbon. Given the absence of any legal framework in PNG for asserting property claims to forest carbon or its sale, the media and spectacle became an important way of creating, representing and also destroying contractual-like relationships that took the place of more formal ways of making people accountable to each other.[1]

Stories from Wawoi Falls about the Origins of Carbon Trading in the Kamula Doso Region

Many Kamula people apparently first became aware of a market for carbon when a group of businessmen and supporters visited Wawoi Falls during the 2005 Independence Day celebrations. Wawoi Falls is a place with a good airstrip where a number of landowners of the Kamula Doso concession reside. Among the visitors were a large American and the son of then prime minister Michael Somare. In addition there may have been someone called 'Steve Dore' (a possible reference to the company Independent Timber and Stevedoring (IT&S), discussed below).[2] According to one story, the party arrived in a helicopter owned by the former prime minister Sir Julius Chan.

Wisapiye Susupiye, a director of the landowner company Tumu Timbers Development Pty Ltd (commonly known as Tumu Timbers), also arrived and talked about a carbon sink company and how this business may be difficult to organise because the World Bank had left PNG. Despite the departure of this powerful bank,[3] some Kamula from Wawoi Falls told me that the presence of the American — a 'white skin' from a country more powerful than Australia — was an excellent feature of the proposal. They made the point that the 'ways' of Australians and Americans were far better than the Malaysian–Chinese management of the Wawoi Guavi logging concession. According to some Kamula, the American, and his supporters like Susupiye, had indicated that everyone working on the project would eat in the same mess. This was very different to the situation at Kamusi, the headquarters of the Wawoi Guavi concession, where only expatriates and a very small and select group of nationals could eat together. There was also talk of how bank accounts were to be opened in Australia for landowners, of creating good roads and putting 'medicine' in the rivers to ensure there was no pollution. Susupiye and the American reported that

---

1   Some of this phrasing derives from Fay (2013).
2   A United States company, Independent Timber and Stevedoring, had about one third of the shares in IT&S. But in April 2010, Hilo Investments, an Australian company, acquired 7 million shares in PNG IT&S. Who is behind Hilo is currently unclear.
3   For some Kamula the World Bank might be linked to One World Government — an apocalyptic and devilish political regime that could precede the return of Christ.

the minister of forests had been paid K6 million for supporting the extension of the Wawoi Guavi into the Kamula Doso area. They also said that now there was no support within government for Rimbunan Hijau (the owners of the Wawoi Guavi concession) to be given rights to Kamula Doso's timber via the extension of the Wawoi Guavi concession.[4] At the time this statement was made there were in place a series of legal challenges to the decision by the PNG Forest Authority to authorise this extension.[5] While the outcome of these challenges was not yet known, people at Wawoi Falls understood that the American was proposing a useful alternative to logging by Rimbunan Hijau, a company whose track record in the Wawoi Guavi concession was well known at Wawoi Falls and a frequent source of complaint. People also understood that Steve Dore's alternative proposal had something to do with carbon trade. But over the next two years no new developments of this carbon proposal were reported. And it later became clear that IT&S was mainly interested in linking large-scale logging with road building and the acquisition of long-term leases over vast amounts of customary land in Western Province (Mirou 2013: 385–531).

Moves to create commodified forest carbon in PNG as a potential alternative to logging were spurred by global policies promoting REDD (reducing emissions from deforestation and forest degradation) schemes. However, between 2000 and 2005 a series of subjudicial reports on the mismanagement of PNG's forest sector highlighted major difficulties in government regulation of the sector (Ombudsman Commission of PNG 2002, 2004; The 2003/2004 Review Team 2004a, 2004b, 2004c, 2004d). Evidence also emerged that accessible primary forests might be logged out within 20–30 years (Filer et al. 2009; Shearman et al. 2009). If the government could not regulate the forestry sector on a sustainable basis it was unlikely to be able to implement policies that would create effective REDD schemes (Melick 2010).

A great deal of evidence suggesting that PNG was unprepared for carbon trading came from the activities of Nupan Pty Ltd during the period 2007 until 2010. This company was led by Kirk Roberts, who was interested in developing carbon projects throughout PNG, including Kamula Doso. Nupan seems in 2007 to have established a working relationship with Mr Kond, of Koo Management. Mr Kond had apparently influential connections via his position as vice president of Prime Minister Somare's political party. Kond's role was to 'liaise with and advise the PNG government' on Nupan's behalf to ensure Nupan

---

4   For background on the extension of Wawoi Guavi into Kamula Doso see the Ombudsman Commission of PNG's (2002, 2004) reports and that by the PNG Forestry Review Team (2001). For details on the Wawoi Guavi concession see Independent Forestry Review Team (2001, 2003) and the reports of the 2003/2004 Review Team (2004a, 2004b, 2004c, 2004d).
5   And in July 2010 the National Court found that the Forest Management Agreement between the landowners and the state was invalid. This agreement gave the state rights to 'manage' the timber in the Kamula Doso concession.

had access to sites containing carbon. Nupan also established an arrangement with an Australian-based company — Carbon Planet — that would broker the carbon credits to the global markets.[6]

Figure 9.1 outlines how Nupan mapped its proposed Kamula Doso project. While there are problems with the locations of places, the map broadly indicates that the proposed carbon project completely overlapped with the proposed Kamula Doso logging concession.

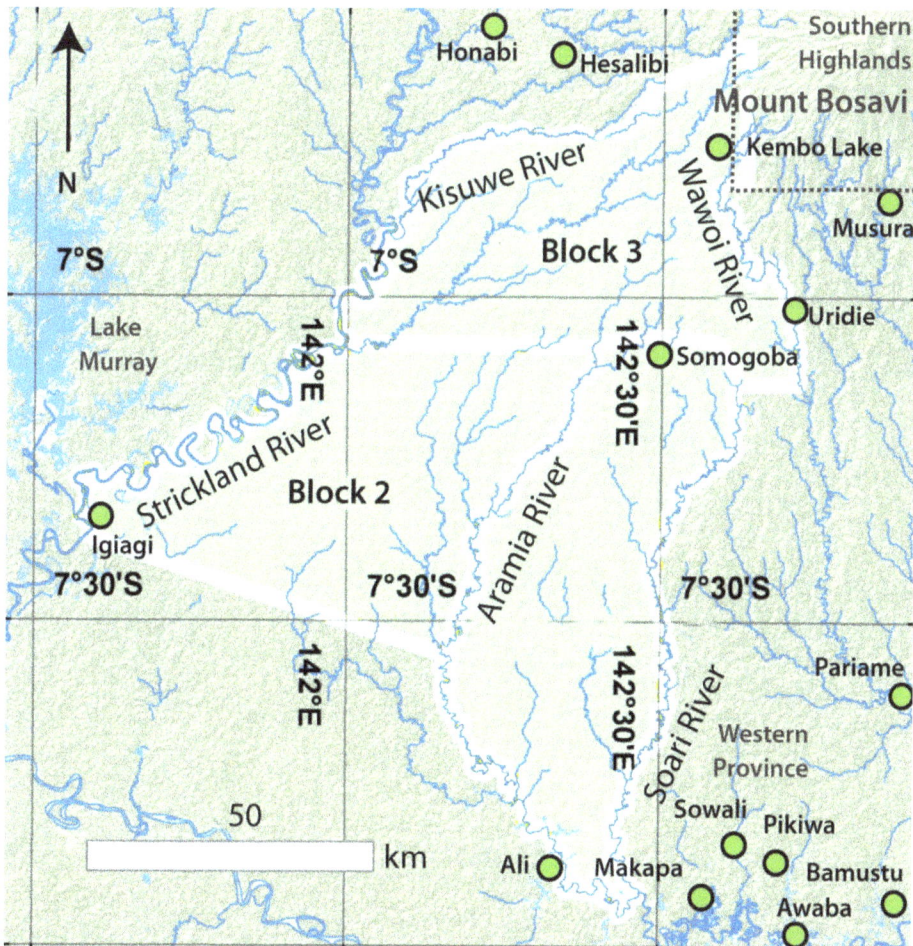

Figure 9.1 Nupan's version of the Kamula Doso concession.

Source: Tumu Timbers Development Limited (2010: 21).

---

6   Viewed at www.carbonoffsetsdaily.com/news-channels/global/broker-paid-k200000-for-carbon-%E2%80%98consultancy%E2%80%99-15695.htm. In April 2008 another company, Forest Top, also became involved in attempts by Carbon Planet, Nupan and Koo Investments to develop carbon projects in PNG (personal communication, Ilya Gridneff, Port Moresby, 16 July 2009). I was for a brief period a consultant to Carbon Planet (Wood 2009).

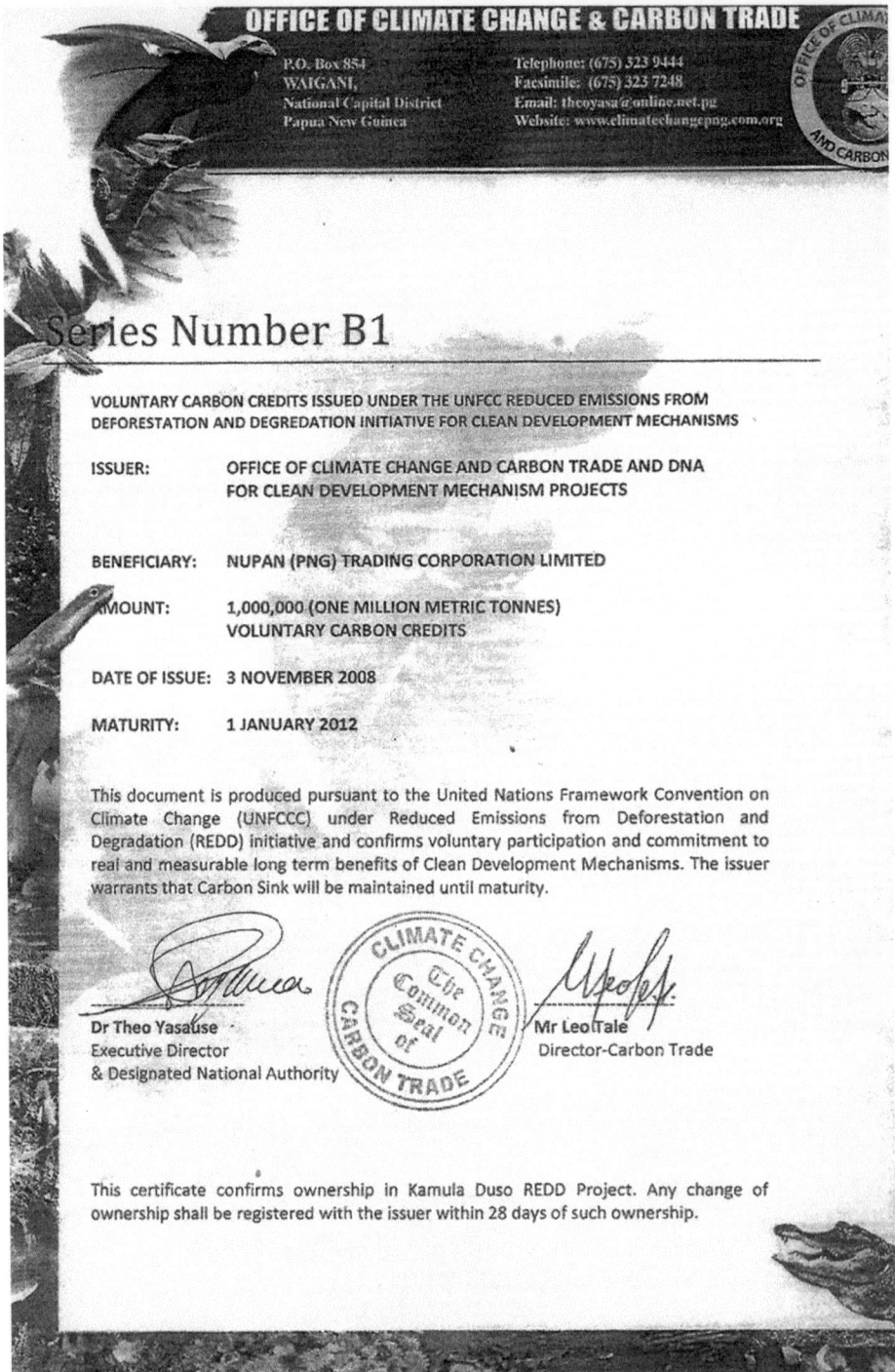

OFFICE OF CLIMATE CHANGE & CARBON TRADE

P.O. Box 854
WAIGANI,
National Capital District
Papua New Guinea

Telephone: (675) 323 9444
Facsimile: (675) 323 7248
Email: theoyasa@online.net.pg
Website: www.climatechangepng.com.org

Series Number B1

VOLUNTARY CARBON CREDITS ISSUED UNDER THE UNFCC REDUCED EMISSIONS FROM
DEFORESTATION AND DEGREDATION INITIATIVE FOR CLEAN DEVELOPMENT MECHANISMS

ISSUER: OFFICE OF CLIMATE CHANGE AND CARBON TRADE AND DNA
FOR CLEAN DEVELOPMENT MECHANISM PROJECTS

BENEFICIARY: NUPAN (PNG) TRADING CORPORATION LIMITED

AMOUNT: 1,000,000 (ONE MILLION METRIC TONNES)
VOLUNTARY CARBON CREDITS

DATE OF ISSUE: 3 NOVEMBER 2008

MATURITY: 1 JANUARY 2012

This document is produced pursuant to the United Nations Framework Convention on
Climate Change (UNFCCC) under Reduced Emissions from Deforestation and
Degradation (REDD) initiative and confirms voluntary participation and commitment to
real and measurable long term benefits of Clean Development Mechanisms. The issuer
warrants that Carbon Sink will be maintained until maturity.

Dr Theo Yasause
Executive Director
& Designated National Authority

Mr Leo Tale
Director-Carbon Trade

This certificate confirms ownership in Kamula Duso REDD Project. Any change of
ownership shall be registered with the issuer within 28 days of such ownership.

Figure 9.2 Carbon credits from Kamula Doso issued by the Office
of Climate Change for the benefit of Nupan Pty Ltd.

Source: PNG Office of Climate Change and Carbon Trade.

During 2008 something like a working relationship developed between Nupan, Carbon Planet and the PNG Office of Climate Change (OCC).[7] This culminated on 3 November 2008. Theo Yasause, director of the OCC, signed a certificate that represented a million tonnes of carbon credits derived from the Kamula Doso concession (Figure 9.2). The certificate gave the beneficiary — Nupan — ownership of these credits that were said to 'mature' on 1 January 2012. In retrospect this was probably the high point of that relationship and the beginning of the end of any serious likelihood that Nupan's Kamula Doso carbon project could be realised.

While the credits were called 'voluntary carbon credits', they were also said to derive from the 'United Nations Framework Convention on Climate Change (UNFCCC) under Reduced Emissions from Deforestation and Degradation (REDD) initiative'. However, at the time the certificate was signed, no such UN-authorised REDD initiative existed and nor did any UN-regulated market for REDD carbon credits exist. But a post-Copenhagen version of such a market was expected to have emerged by the 'maturity' date.[8] The certificate was a claim to the existence of carbon credits and implied that Nupan already had rights of ownership in such credits. The certificate was also a promise that in the future — on maturity — REDD-compliant credits would exist. But since the certificate also mentioned voluntary credits it seems Roberts was allocated rights to sell into this market as well. The credits were neither authorised, nor constrained, by any specific state legislation regulating their definition or sale. Their creation was possibly just rather silly. But given the certificate's elaborate design it may have been a spectacularly naive attempt to assert state authorisation or even a deliberate attempt to defraud — assuming anyone could be found to take the document seriously.

As a consequence, it has proved hard for anyone to provide a rational explanation of why the certificate was produced beyond the 'capture' of Nupan and OCC officials and the thrill of the spectacle they were creating. Jim Johnson, of Carbon Planet, used talk of symbols to explain the documents as something less than real, as standing for something else beyond the money they appeared to contain:

> … there are in existence a group of certificates issued by OCC to Nupan for Kamula Doso. They are not real certificates. They are symbolic to the fact that Nupan is recognised by OCC as rightful developer to the landowners of Kamula Doso. (natashaloder.blogspot.com, 6 June 2009)

---

7   This office was also known, at various times, as the Office of Climate Change and Carbon Trade and the Office of Climate Change and Environmental Sustainability. In this chapter I will refer to it as the Office of Climate Change (OCC).

8   Loder has noted that 14 per cent of the forest carbon traded on voluntary markets in 2009 was based on promises to deliver REDD credits (natashaloder.blogspot.com, 6 June 2009).

Dr Yasause also adopted a similar response. He presented the credits as representing a provisional copy of an original that perhaps did not yet exist. He implied that this 'original' could be brought into existence after proper consultations with a newly unified group of landowners:

> Kirk Roberts of Nupan PNG had been working with the landowners of Kamula Doso and made representation to my Office. We issued an Interim copy of the Certificate dated 3 Nov 08, on the basis that they will undertake wide stakeholder consultation covering all landowners, provincial and local government before such project can be entertained. No documentation was submitted with proof that consultations were undertaken and the people are happy with possible carbon trade … Dr. Theo Yasause … has since became aware that not all landowners were involved in or supportive of the project being developed by Nupan. Since 3 February 2009 the OCC ceased to have any direct links with Nupan. (natashaloder.blogspot.com, 6 June 2009)

Even if little more than a speculative claim about some unlikely future state of the world, the possibility of carbon credits did help convey powerful material effects to some involved in their production. For his work for Nupan and Carbon Planet, in May 2008 James Kond was apparently paid K200,000 (A$85,000) by Carbon Planet. In addition, Carbon Planet paid some money to Kirk Roberts and his company Nupan (Ilya Gridneff, 24 September 2009, Australian Associated Press).[9] There was also some expectation that Carbon Planet would help the OCC develop regulatory policies and procedures. Further reiterating the value of these highly speculative transactions, in July 2009 Carbon Planet proposed a merger with an Australian company called 'm2m Group'. Carbon Planet claimed to have some 25 potential carbon trading projects in PNG that could generate A$1 billion a year (ibid.).[10] While Carbon Planet made useful claims about its assets, Dr Yasause was sacked partly for issuing 'fake' credits for Kamula Doso and other carbon projects in PNG, and the OCC was moved from the prime minister's office to the Department of Environment. As a result there was no immediate speculative boom in carbon credits derived from Kamula Doso, or from PNG generally.

Media Accounts and Rimbunan Hijau

The creation of the credits was subject to intense media scrutiny and critique. These stories positioned Kirk Roberts as a 'colourful character', 'carbon cowboy', 'carbon conman' and 'larrikin'. These terms, and associated stories of his earlier troubles with the law concerning doping of horses, suggested he was duping the

---

9   Gridneff's report is available at probeinternational.org/library/wp-content/uploads/2011/11/More-bad-news-from-Papua-New-Guinea-_-redd-monitor.pdf (viewed 8 May 2015).
10   This merger collapsed and Carbon Planet itself became bankrupt.

people living in the Kamula Doso area. There was virtually no attempt to represent Kamula, or other land users, except as victims of a scam (and, more recently, as victims of coercion). Kirk Roberts became a useful representative of other carbon brokers operating in PNG and a broader signifier of what can go wrong with the commodification of ecological services through REDD schemes and carbon trading. The Nupan project indicated that attempts to construct an unfeasible market in carbon credits was, in PNG (and other parts of the world), likely to be intrinsically corrupt and corrupting (Lohmann 2009). The Nupan project also highlighted the possibility that capitalism, in general, or more specifically in Third World resource frontiers, deployed deception and spectacle as it its normal way of operating.

The dominant impact of the media was to deny the Kamula Doso carbon project any legitimacy and make it appear as an exemplar of the inability of a corrupt PNG government to effectively govern the new resource. In the lead-up to Copenhagen with intense debate about various proposals concerning REDD schemes (see Gabriel, Chapter 10, this volume), these stories about PNG's carbon credit schemes, particularly concerning Kamula Doso, briefly gained some international significance.

The media played down the role of pre-existing timber interests which were also criticising Nupan's carbon project. One example of intervention by such interests emerged in January 2009 when the 'Kamula Doso Landowners' group delivered a petition to the forests minister (then Belden Namah) that opposed any carbon trading and supported logging. A version of the petition was also apparently published in *The National* newspaper, owned by Rimbunan Hijau. Around the same time the petition was presented to the government, a lawyer working for Rimbunan Hijau had provided remarkably good advice about carbon credit schemes to the Kamula, especially to those Kamula who strongly supported Rimbunan Hijau. Copies of this advice, along with copies of media coverage about the OCC issuing fake carbon credits to Nupan, were circulating at Wawoi Falls when I stayed there in June 2009.

While there I came across a draft version of perhaps the same petition that was presented to the forests minister. The content partly derived from Rimbunan Hijau's opposition to Nupan, but other ideas found in the draft derived from, and alluded to, local concerns about international governance (and the apocalyptic advent of Christ).[11] The purpose of the petition was to prevent:

---

11   Rimbunan Hijau may be one source of this kind of thinking. Its newsletter of October 2009 has the headline 'Climate Change: One-World Government Conspiracy' over an article outlining Lord Monckton's views on climate change (RH PNG 2010). But the petition writers may have also been thinking of local apocalyptic political theologies where a warming world, corruption and the appropriation of 'good wind' by Europeans are not radically unexpected phenomena. When Wisapiye Susupiye visited Lake Campbell with Kirk Roberts he found that some people there were quite explicitly concerned that the carbon project 'will lead to the emergence of new money. People were thinking it is something to do with 666 and One World Government. This is because they are uneducated' (from notes of a conversation in English with Wisapiye Susupiye, 15 December 2009). This kind of thinking was also reflected in reports from people at Lake Campbell that while visiting there Kirk Roberts had clearly displayed a K1 million note.

The proposed establishment of International Carbon Trade Network in Papua New Guinea and one of which is Kamula Doso FMA [Forest Management Agreement]. We believe the National Government may have endorsed the Carbon Trade Network internationally deliberately for two existing logging projects in Papua New Guinea without prior consent of customary landowners ... What the government has done is absolutely erroneous and basically [a] deprivation of rights in management of our forests as rightful resource owners. ... Our custom, tradition and legal rights must never be wrongfully motivated and manipulated by way of corruption and illegal engagement of foreign organization such as International Carbon Trade Network ... (Kamula Doso Landowners, 19 January 2009. Slightly corrected.)

The petitions, as signs of Rimbunan Hijau's public concern, combined with the revelations of the 'fake' carbon credits, and indications (discussed below) of landowner conflicts over the control of the landowner company Tumu Timbers, added to the withdrawal of support from the OCC for Nupan's project as a REDD demonstration scheme.

Nupan and Carbon Planet then started to place more emphasis on the Kamula Doso carbon scheme as linked to the voluntary market, rather than as part of any UN-sponsored transaction. But later, in August 2009, Dr Wari Iamo, acting director of the OCC, seemed to close down the voluntary market as a serious option:

Voluntary Carbon Agreements (VCA) are not currently supported by the Government as they fall outside formal REDD arrangements and are highly unlikely to deliver high carbon prices. PNG would be selling its forests cheaply by going down the VCA path. (*The National* 31 August 2009)[12]

Despite this setback, during 2009, the likelihood that Kamula Doso would be logged also diminished. Eco-Forestry Forum, in court over attempts dating back to 1999 by the PNG Forestry Board to treat the Kamula Doso logging concession as an extension of the neighbouring Wawoi Guavi concession owned by Rimbunan Hijau, secured a court injunction against the PNG Forest Authority from undertaking any activities related to Kamula Doso pending the outcome of the legal proceedings. In June 2009, Eco-Forestry Forum secured a further order from the National Court to specifically stop the OCC from taking any further

---

12   However, Nupan's capacity to claim some degree of government authorisation for trading credits on the voluntary market position improved within a month when Pius Ripason of the OCC wrote a letter to a strong supporter of Nupan. In this letter, dated 21 September 2009 and displayed on Nupan's website, Ripason stated:

As for the voluntary market, we have no control over this venture because OCCES has no legislation or law to cover voluntary market on customary land. Consent to participate in [the] voluntary market on customarily land can be arranged or agreed upon between resource owners and developers which ever suits your ILG and the developer. It is the resource owners or ILGs who grant permission/ consent to the chairman of the ILG or some one of their nominee to act on their behalf to consult with a developer.

steps to issue rights over the forests of Kamula Doso. The Department of Lands was also given an injunction. In April 2009 the Department of Lands gazetted a 99-year Special Agricultural and Business Lease on behalf of the landowner company Tumu Timbers, over all the land in the Kamula Doso concession area. The lease, organised by IT&S rather than Nupan, allows for virtually any activity on the land and could constitute a way of engaging in forestry by avoiding the requirements of the *Forestry Act 1991*. It makes the control of Tumu Timbers a very valuable resource for any developer interested in the Kamula Doso area.

Nupan's Use of Media

Along with attempting to maintain its exclusive control of Tumu Timbers, Nupan also had to respond to mainstream media criticism of the fake carbon certificates and the project generally. During 2009 to 2010, Nupan placed large amounts of material on its website.[13] Kirk Roberts, representing Nupan, intervened in internet discussions and made himself, and his supporters, available for numerous interviews. Officials from Carbon Planet were also active on the internet in responding to criticism of their involvement in PNG's carbon credit schemes.[14] Roberts used his website to demonstrate the existence of contracts and agreements between himself and his supporters. For each project he was promoting in PNG, landowner groups who supported his carbon projects received a variant of the following kind of letter:

Mr. Wisa Susupe, July, 2009

KAMULA DOSO FMA Block 1;2;&3 FMA

Welcome Sir!

It is our great pleasure to welcome you and acknowledge your fellow Vice Chairman and Directors of Tumu Timbers Development Limited ...

We acknowledge receipt of your Board Minutes confirming our appointment, and assure you that we have, on your behalf, and with the help of the People of Kamula Doso, now completed all the Corporate, Legal, Government, and Social responsibilities required of us as your appointed Power of Attorney, and that we will at all times continue to act in your best interests.

... We confirm that the independent verification process to enable your Project to be formally recognized under the UNFCCC guidelines for REDD Carbon Credits is now well underway.

13   This website, www.carbonowontok.org, has since been removed from the internet.
14   Viewed at www.carbonplanet.com/REDD_addressing_the_issues; www.thepunch.com.au/articles/medias-carbon-confusion-is-grist-for-the-lumber-mill/. These websites have been removed from the internet.

The decision you have all made to preserve your beautiful forests from logging and other destructive activities is a brave one, and we salute your intention to maintain your Forests in the interests of providing Mother Earth with 'A breath of fresh air'.

We salute you and thank you for your trust.

Yours sincerely,

'Roberts'
Kirk William Roberts
CEO / Chairman
Nupan (PNG) Trading Corporation Limited.[15]

Along with posting on his website such written expressions of agreement, Roberts also posted video recordings of village-based rituals of welcome and talk (the latter usually by Roberts), but sometimes covering local expressions of support. In these videos local support was also presented through an emphasis on images of local 'traditional' ritual (which at Lake Campbell was paid for by Roberts). This emphasis on local culture deployed a style of representation similar to that used by Greenpeace (2004a, 2004b), which often portrays people from the Kamula Doso region in traditional dress, opposed to logging, whereas Roberts presents tradition as partly constituting an authentic agreement to the carbon project while also suggesting that the carbon scheme could conserve culture as well as rainforests. It seems that Roberts regarded these rituals of agreement, and their circulation into the global media via Nupan's website, as more than epiphenomenal to the 'real' politics of carbon and considered them as equally crucial sites for the creation of carbon markets. Roberts's more formal relations with state officials, landowners and companies were secured through texts vesting power of attorney in Roberts and in letters of support. However, while Greenpeace imagery rarely includes images of its activists, Roberts was presented in the Nupan videos as the key focus of the event, often becoming part of the tradition, while at the same time being sufficiently different to act as a powerful embodiment of local interest and support for the carbon project.

Video recording landowner agreement to carbon trading was sometimes explicitly defined in reference to a larger, often global, audience. According to Roberts, the landowners, especially if they agreed to carbon trading, would apparently become global celebrities. But the kind of effect often created in the video actually reduced that global significance to something more ephemeral, like 'fame'. In a video, for a time placed on his website, Kirk Roberts is shown explaining to residents of Somokopa how such global fame was one benefit of his carbon credit scheme. He said:

---

15   www.redd-monitor.org/wp-content/uploads/2009/09/kmdso-july.pdf (viewed 8 May 2015).

It's very, very important that we do not let logging come in and kill all the forests and your beautiful land. We have been travelling around all the villages in the Western Province and in the Kamula Doso district. Just to let you know what we have been doing … Now we are getting very serious. Very close to finishing. Everybody in the world is very interested. These cameras are now recording this story and these meeting places to put on an international website for the whole world to see. All your faces will be in the world. In the world's eyes. And everybody will know this village. And know your support in regards to carbon trading. And you will be famous (applause).[16]

But as this transcript suggests, some of the videos — especially those concerning Kamula Doso — mainly involve recording Roberts as the central figure in the story rather than the residents. What the Kamula Doso videos best record was Roberts himself talking and, at times, explaining his own ideas about global fame. His own centrality in the videos undercuts any easy understanding of landowner support.

The emphasis on Roberts may have derived from his video team adopting certain conventions and content evident in mainstream conservation media productions (Igoe 2010; Igoe et al. 2010). According to Igoe and colleagues (2010: 493), these productions 'portray celebrities, corporate leaders, and high profile conservationists as a heroic vanguard in the struggle to save the planet'. The idea that Roberts might be a heroic vanguard is made explicit in *Evolution*, a professional film made for Carbono Wontok and Nupan.[17]

Evolution is a composite of all the videos taken at the sites of Nupan's various carbon projects. The voice-over presents Kirk Roberts as an 'infamous' hero in the struggle against logging:

Right now just one man and a whole community stands between the rapacious loggers and the rainforest. His name is Roberts and in recent times he has become infamous for the relatively innocent activity that the ignorant and ill informed mainstream media have taken great delight in blowing out of all proportion. But Roberts understands the power of the people to decide their own future. And for several years he has been trying to educate and inform landowners village by village of the potential to save their rainforest and helping to save the world.[18]

Whatever heroic attributes were attached to Roberts by this film, they were perhaps limited by the unique understanding of PNG conveyed to the audience in the introduction to the film:

This is the story of a mystical country you probably have never heard of. If you have heard of it chances are you do not know exactly where it is. It is the story

---

16 vimeo.com/9149934 (viewed 20 January 2015).
17 The film was made by Lauren Horner and Morgan Smith with 'concept' by Jessica May Lyla.
18 vimeo.com/9617496 (viewed 20 January 2015).

of a culture that started over 9000 years ago and throughout all that time it has managed to keep its identity and culture intact. Against all the odds in spite [of] modernization, commercialization and computerization, PNG has managed to maintain its traditional values — the family, the village, the clan … Even when they venture out to the relative civilization of their major cities they take their culture with them in the form of *wontok*. *Wontok* means one talk in pidgin.[19]

The influence of Nupan's media productions was also limited by the fact that they did not really attract much of an audience. One internet site where the Nupan videos are stored indicated the number of viewings of some of those videos by late 2011 (Table 9.1).

The last three videos in Table 9.1 relate to the Kamula Doso area. The number of plays indicate that very few people watched these videos.[20] So these attempts to use media failed to create much global celebrity status for Roberts, his supporters, or for Nupan. But direct uploading from the Nupan website is not the only way of viewing Roberts' videos. In early December 2009 Al Jazeera's program *People & Power* ran a story on 'Carbon Cowboys' in PNG. It contains many excerpts from videos made by Nupan and a copy on YouTube registered 4,079 viewings.[21] It seems unlikely that such numbers made Roberts or the landowners 'globally' famous.

Table 9.1 Number of viewings of selected Nupan media productions, 2011.

Name of video	Date uploaded	Number of plays	URL
Evolution	21 Feb 2010	131	vimeo.com/9617496
ILG Signing Ceremony (Part 1)	2 Feb 2010	21	vimeo.com/9056479
ILG Signing Ceremony (Part 3)	31 Feb 2010	33	vimeo.com/9120963
ILG Signing Ceremony (Part 4)	1 Feb 2010	19	vimeo.com/9122721
ILG Signing Ceremony (Part 5)	2 Feb 2010	21	vimeo.com/9149934
Kirk Roberts – Somongapa Village – 8 October 2009	30 Oct 2009	12	vimeo.com/7343655
Lake Campbell Sing Sing – 8 October 2009	8 Nov 2009	14	vimeo.com/7498955
Somongapa Sing Sing – 8 October 2009	5 Nov 2009	16	vimeo.com/7464175

Source: YouTube.

Roberts also used print journalists to promote his case for the development of carbon sinks on the voluntary market. In October 2009 a journalist with PNG's *Post-Courier* visited the Kamula Doso area with Roberts, a video crew,

---

19   vimeo.com/9617496 (viewed 20 January 2015).
20   My own repeated viewings may be counted here.
21   Viewed at www.youtube.com/watch?v=LCzGIsKEw48&feature=player_embedded. The number of hits on the site was noted on 1 November 2011.

and a scientist who was presented as an international verifier of carbon credits.[22] The journalist offered readers a vision of the landscape that might be equivalent to a print media wide shot. Typically taken from the air, the long shot involves stretches of nature 'so vast and exotic that they cannot fail to suggest unlimited possibilities' (Nugent cited in Igoe 2010: 382). But the journalist's description suggests he was in the highlands rather than the lowlands of Western Province:

> Over the next two days we visited three small tribal villages in the Kamula Doso Project area — Awaba, Lake Campbell, and Somogopa, only accessible by light aircraft, and then only on a good day weather wise. The mountains tower above five thousand meters in every direction and the clouds are often sitting just a few hundred metres above the ground, making for some interesting flying conditions![23]

The mountains are actually 20 kilometres away from Lake Campbell and some 80 kilometres from Awaba. Following Tsing (2005: 68) on spectacular accumulation's links to resource frontiers, we could say the journalist's portrayal asked his readers to see a magical landscape that did not exist — or at least not in the way he implied.

The journalist went on to paint a vivid picture of landowner support for the Nupan project in the rather flat country surrounding the three places they visited. In the process he creates some enthusiastic, and extraordinarily numerous, 'local people':

> The obvious question we all had was how much did the local people know about Carbon Trading, and we were constantly surprised by both the quality and depth of the answers, and the penetrating questions we were asked in turn.

> What did we think about the village; why did the media attack PNG so much; and why is the Carbon Trading Project taking so much time.

> … Lake Campbell was a beautiful location, this time a thousand people had walked in for the ceremony, and the welcome and positive reinforcement for Carbon Trading was evident.[24]

---

22   probeinternational.org/library/wp-content/uploads/2011/11/More-bad-news-from-Papua-New-Guinea-_-redd-monitor.pdf (viewed 8 May 2015).
23   probeinternational.org/library/wp-content/uploads/2011/11/More-bad-news-from-Papua-New-Guinea-_-redd-monitor.pdf (viewed 8 May 2015).
24   probeinternational.org/library/wp-content/uploads/2011/11/More-bad-news-from-Papua-New-Guinea-_-redd-monitor.pdf (viewed 8 May 2015).

Stories of Competing Kamula Links with Developers and More Media Excess

While Nupan was attempting to create these spectacular images of landowner support for the carbon project, its primary supporters — the executive of Tumu Timbers — fragmented. This fragmentation was due to opportunities provided by IT&S. While IT&S probably lacked the expertise and capital itself to undertake logging in PNG, it had since the mid-2000s an interest in the development of the Kamula Doso area. During this time, one aim of IT&S was to develop a road (the 'Trans-Papuan Highway') that would run from around Kiunga to just south of Nomad, and continue south towards Balimo and Kawito. The intention was to undertake logging in areas adjacent to the road. The logging was to be organised through Timber Authorities. By 2007, IT&S had formalised the interest of the executive of Tumu Timbers in such a project and the two companies agreed to create a joint venture company known as Pisa American Lumber Ltd.[25]

In 2008 Tumu Timbers started dealing with Nupan and in late October agreed to appoint Nupan as preferred developer of a carbon project in Kamula Doso. In the process of making this agreement, Tumu Timbers vested power of attorney in Kirk Roberts. However, IT&S quickly re-established working relationships with some board members of Tumu Timbers. In November 2008, the board of directors of Tumu Timbers revoked its agreement with Nupan, and in January 2009 Tumu Timbers board members also rejected dealings with Carbon Planet.

But by March 2009 some of the Tumu Timber directors resumed their relationship with Nupan. By then, chairman Wisapiye Susupiye and other directors accompanied Kirk Roberts, James Kond and representatives from Carbon Planet on a trip to the Kamula Doso area to gain the endorsement of landowners. On the same day the team visited Awaba and then Wawoi Falls. According to my sources at both places, relevant incorporated land group chairmen were paid K500 to sign a document that apparently endorsed Nupan as the preferred developer of Kamula Doso as a carbon project. At Wawoi Falls a more public meeting was held and again most incorporated land group chairmen signed this document. However, at Wawoi Falls they came under sustained attack from supporters of Rimbunan Hijau who argued against choosing carbon as a development option.

---

25   Pisa is the name of a mountain in Kamula territory. Its conjunction with 'America' suggests what many landowners would regard as a highly desirable alliance between powerful Europeans and the Kamula. While Tumu Timbers secured the lease–leaseback agreement over the project area (Kamula Doso) it was also obliged to lease to Pisa 'the whole of the Project Area for the purpose of logging and taking of timber and associated purposes' (joint venture agreement between Tumu Timber and IT&S, cited in Mirou 2013: 523).

They argued that the proposed carbon trade had no legislation regulating its operation, and that middlemen such as Roberts, Kond and Susupiye would be the main beneficiaries.

Despite criticisms of carbon development from people at Wawoi Falls, and despite the fact that Susupiye never returned to Wawoi Falls, the documented support gained from his visit there, and to other villages, enabled him to hold a meeting of Tumu Timbers in Port Moresby that officially approved changing Tumu Timbers' development strategy from its previous approval of the 'FMA and Agriculture TA' (when it supported IT&S) to a carbon credit project.[26] The meeting also re-endorsed Kirk Roberts as the developer of the carbon project with Koo Management endorsed as PNG project coordinator for Nupan.

The significance of these continual shifts in who apparently had control of Tumu Timbers increased dramatically once people became aware of a notice published in the *National Gazette* at the end of April 2009. This notice advised that Tumu Timbers had been granted a 99-year Special Agricultural and Business Lease over the 790,800 hectares making up the Kamula Doso timber concession.[27] Perhaps reflecting the new importance of controlling the lease owned by Tumu Timbers, on 6 June 2009 Abiliye Wape, who was a supporter of IT&S, was voted out of his role as acting chairman of Tumu Timbers. He and some of his supporters were charged by Nupan supporters with the 'misconduct' of maintaining contact with IT&S.

Further highlighting the importance of controlling Tumu Timbers and its special lease, on 18 June 2009 Wape and his supporters held their own meeting of Tumu Timbers at which they removed Susupiye and his vice chairman from the board. These changes, which made Wape chairman, were lodged with PNG's registrar of companies, the Investment Promotion Authority, and then, so the story goes, the file was 'restricted' to prevent the Nupan supporters from gaining access to knowledge about their loss of control of Tumu Timbers. In addition, the supporters of IT&S gained a court order preventing prior chairman Susupiye from representing Tumu Timbers. The Nupan-supporting version of Tumu Timbers responded to Wape's court injunction by asking the police fraud squad to investigate the possibility of charging Wape with fraud over the legality of his Tumu Timbers meeting. They also referred the IT&S lawyer to the PNG Law Society for misconduct as he had apparently been instrumental in calling the meeting of the Tumu Timbers Board that had reinstated Wape as chairman. And they apparently secured the support of the Investment Promotion Authority,

---

26 Tumu Timbers Minutes of Special Meeting, held at Holiday Inn, Port Moresby, 6 June 2009. Copy in author's possession.

27 This was one of many such leases that have been granted since the Somare government came to power in 2002, and that now cover more than 5 million hectares of land (Filer 2011), but the lease to Tumu Timbers was and still remains one of the largest.

which joined Susupiye as a co-defendant in the court hearings concerning Wape's injunction against Susupiye. Apparently they did this because Wape had misrepresented to them the nature of the meeting that had reinstated him as chairman of Tumu Timbers. In June 2009 two boards of Tumu Timbers existed — one supporting IT&S and the other supporting Nupan. The disputing boards of Tumu Timbers then entered into court proceedings against each other.

As part of this ongoing dispute, in December 2009 Wape appeared on SBS television claiming to have been coerced by police into signing documents beneficial to Nupan. The story generated significant interest and Wape became an 'Indigenous leader' of the 'Kamula Doso Peoples' (Lang 2010). Wape's story, in one case headlined as 'Kidnapping Carbon Rights', also became a story of carbon trading's links with human rights violations. OneWorld UK carried the following headlines:

> Carbon trading 'a crime against humanity'
>
> Carbon Markets Violate Indigenous Peoples' Rights and Threaten Cultural Survival
>
> Indigenous Leader Kidnapped and Forced at Gunpoint to Surrender Carbon Rights for REDD in Papua New Guinea. (Lang 2010)

But around the time the story of Wape's coerced consent first appeared on SBS, Nupan's position strengthened. This culminated in December 2009 when mediation sessions between representatives from IT&S, Nupan, the Investment Promotion Authority, and Susupiye and Wape were held. During these sessions I was told by Nupan supporters that Wape made it known he would be prepared to withdraw his legal case if he was paid K6 million from Carbon Planet and Nupan. Around this time Nupan supporters also understood that Nupan would have access to 'A$20 million' as 'mobilisation funds' for the cost associated with the establishment of the Kamula Doso project. Wape's request was not agreed to in the discussions, but the supporters of IT&S apparently withdrew the court action against Susupiye and the other Nupan-supporting board members. According to the Nupan supporters, Wape was paid K4,000 as a gesture of compensation and compassion.

In July 2010 Wape, by now far more supportive of Nupan's carbon project, denied that he was 'taken at gun point' to support Nupan's carbon trade project in the previous year. Instead, Wape said he was bribed to make this claim on the SBS television program:

> Mr Wape was forced to issue the denial after the United Nations Convention on Climate Change (UNFCCC) brought up the matter about the SBS television program when the Kamulo Doso project was undergoing verification under the voluntary carbon standards [VCS] and climate, community and biodiversity alliance [CCBA]. 'What happened,' he said, 'is that I requested a police escort

to the car, because of unrest in the village caused by loggers. I have already apologised to my village and to the board of Tumu Timbers some months ago, and for the UNFCCC to bring this up again at this time, in an attempt to create unrest, is both mischievous and irresponsible,' Mr Wape said. He said as a director of Tumu Timbers, he was working towards finishing the first carbon trading scheme in PNG to be dual verified under the VCS/CCBA protocols. Managing director of Tumu Timbers … confirmed the statement by Mr Wape. (*Post-Courier*, 8 July 2010)[28]

End Game and New Theft of Land?

Unfortunately for Nupan the verification process elicited a highly critical response from OCC director Wari Iamo and from the lawyers he commissioned to provide a legal opinion on Nupan's property rights in carbon. Equally critical were a vast array of letters from land groups in the Kamula Doso region.[29] But in September 2010 Nupan did succeed in mobilising public expressions of support from some sections of the PNG government. On 22 September Mr Singin, a legal adviser to the prime minister, indicated a degree of support for Nupan[30] and the next day the then acting prime minister, and deputy prime minister, Don Polye, minister for works and transport, wrote a letter endorsing Nupan:

> … in their endeavours to create and verify the first four (4) commercial voluntary carbon trading projects in PNG — namely Kamula Doso, East Pangia and Aitape-Lumi … and Passismanua Inland forest concession areas … it is our view that the landowners have the inalienable right under customary, commercial and parliamentary laws to manage and do carbon trading project(s) that they may develop from their forest concession areas. Any government direction that relates to the people's forests is suborned by the rights of the people through ILGs to decide that future.[31]

---

28   These easy shifts in Wape's position left some observers surprised and somewhat injured. Brian Thomson, the journalist who had run the initial story of Wape's coerced support on SBS News, tried to maintain the truth of his initial story: 'The landowner concerned, Abilie Wape, came to my hotel willingly, knowing that I was investigating claims that landowners were being put under pressure to hand over the rights to the carbon in their forests. Abilie made the claims without co-ercion, encouragement or payment. It is quite clear to me that Abilie has since been put under pressure by someone to change his story. … I do not, and never have I, offered payment to anyone. The story published in the *Post Courier* is utterly false.' (Viewed at www.redd-monitor. org/2010/09/14/redd-projects-in-papua-new-guinea-legally-untenable/#response).
29   One response was by a man from Lake Murray who had been active in opposing Doso claims of ownership of the land around the gas field located in the Kamula Doso concession. The other letters all have exactly the same content and may have been organised by Rimbunan Hijau. Viewed at climate-standards.org.
30   Originally viewed at www.carbonowontok.org/, but this site has been disbanded. I have not been able to find another source for this material.
31   Originally viewed at www.carbonowontok.org/, but this site has been disbanded.

Despite the ambiguous reference to the people controlling incorporated land groups corrupting government,[32] Rimbunan Hijau apparently felt it needed to respond to what may have been a shift in the government's position. Its newspaper — *The National* — mobilised Kamula landowners supporting Rimbunan Hijau to report:

> Landowners in the Kamula Doso region have reacted angrily to comments made by Deputy Prime Minister Don Polye in support of proposed voluntary carbon scheme (VCS) deals. Paul Sasae, chairman of landowner company Wawoi Temu Holdings Ltd and chairman of the Dawasi ILG, had publicly criticised the proposed carbon deals, saying that PNG's 'carbon cowboys' have acted fraudulently ... 'The carbon project asks us to halt all forestry and agriculture for the next 40 years' ... 'With no way to grow or buy food, what are we supposed to eat?' Sasae said that his community's situation had become more desperate since the cancellation of the Kamula Doso FMA. According to him, the cancellation was driven by a group of Western activists — not the wishes of local communities ... According to Sasae ... the cancellation of the FMA would effectively mean the loss of a potential 1,600 jobs, local annual royalty payments of K8 million and investment worth more than K100 million, as well as improved local infrastructure.[33]

While such a response would not have been unexpected nor particularly troubling to Nupan supporters, far more worrying was that by around this time it appeared that Kirk Roberts was no longer in PNG and had effectively disappeared. Some stories suggested Roberts had acquired funds or commissions — possibly worth US$20 million — that had been deposited into a Hong Kong bank account. By late 2010 Tumu Timbers supporters were looking for other investors such as the Macquarie Bank (2008) and a New Zealand company linked to the husband of Maria Henderson whose mother, from Makapa village, has interests in land within the Kamula Doso area. In February 2011 Tumu Timbers terminated its relationship with Nupan:

> Tumu Timbers Development Limited, a landowner company, terminated the deal after a board meeting found that carbon trade had no legal framework for trade. Chairman Wisa Susupie said they wanted to lease their land out for agro-forestry projects and sustainable logging after learning that carbon trade would never eventuate and that they were misinformed by Kirk William Roberts of Nupan (PNG) Trading Corporation Ltd. ... Mr Susupie said they had no idea what carbon trade was like or if there was a legal framework in place to trade carbon in the world. (*Post-Courier*, 7 February 2011)

---

32 Polye's statement uses the term 'suborn' — to induce a person to commit an unlawful or evil act; to commit perjury or procure testimony — in the place of 'subordinated to' or 'derived from'.

33 www.pina.com.fj/print.php?print=news&o=8038583704cb515add66aaac231599 (viewed 8 May 2015).

The vacuum left by the end of Nupan's carbon scheme was quickly filled by IT&S, whose lawyer visited Wawoi Falls in January 2011. Apparently the lawyer indicated the IT&S project would result in a fully sealed road from Kiunga to Wawoi Falls and from Wawoi Falls to Balimo. A consensus emerged from the meeting at Wawoi Falls to support this proposal while also allowing Rimbunan Hijau to log the remainder of Kamula Doso. This compromise was made because people understood Rimbunan Hijau would never build any permanent roads. The meeting also agreed to the IT&S lawyer's call for a new board to be established for Tumu Timbers and suggested a strong supporter of Rimbunan Hijau should become the chairman. Another option apparently being explored by the PNG Forest Authority was for the landowner companies supporting Rimbunan Hijau and Tumu Timbers to combine and establish an interim board of a new 'mother' landowner company that would have representatives from all existing landowner companies operating in the Kamula Doso area. This company, with advice from the PNG Forest Authority, may pursue the possibility of selectively logging parts of Kamula Doso while also using less viable areas as potential carbon credit schemes.

As these developments were being discussed, IT&S recruited some new Kamula supporters and in May 2011 organised for them to attend a 'signing ceremony' for the 'Trans-Island Highway (stage two) road project agreement' (*The National*, 24 May 2011). The governor-general of PNG, the minister for commerce and industry, and the deputy governor of Western Province also attended. The project involved linking Western Province via the Trans-Island Highway to the Hiritano Highway in Gulf Province that runs to Port Moresby. It was expected that logging would take place in the 'first stage' of constructing the Trans-Island Highway and it is possible that no second stage will eventuate. But given the recent recommendation by the commission of inquiry into Special Agricultural and Business Leases (Mirou 2013: 526, 385–531) that the special leases covering the Kamula Doso area be revoked, then assuming the PNG government enacts these findings, the whole project, like Nupan's carbon credits from the OCC, may turn out to be largely illusory.

Conclusions

This chapter has looked at some of the factors that influenced the way Nupan Pty Ltd tried to assert persuasive property rights in carbon. Despite all the contingencies and conflicts that marked this project, it was part of a broader historical process that has increasingly linked the environment to financial devices and investments in ecological services and values (Igoe 2010: 3; Igoe et al. 2010: 496). Such 'market-based' conservation lays claims to ownership of emergent, partially defined, yet often apparently 'intrinsic', values of the environment (such as carbon), while also 'conserving' these somewhat hypothetical 'polymorphic' values (Lowrey 2008: 70). Such projects seek to augment these values through various additional mediations, such as derivatives trading (that in the Kamula Doso could have been undertaken by the Macquarie Bank), and transform them into 'capital of a more convertible and globally ramifying kind' (Garland 2008: 62).

In different ways Rose (1994), Tsing (2005) and Igoe (2010) have argued that 'spectacle', narrative and drama are crucial to these transformations of natural resources into forms of private property, capital accumulation and value creation. For Tsing (2005: 57), drama is an essential prerequisite of certain forms of accumulation, and Igoe (2010) argues spectacle mediates many forms of legal conservation. But given the absence of legal authorisation of any carbon markets in PNG, Nupan's largely media-mediated project involved a speculative legality and largely fictitious property claims. Nupan was heavily reliant on an array of representational techniques — such as letters of support from government, stories in the print media, and the use of websites and videos to create convincing images of its control of carbon. Nupan also had powerful narratives of deforestation and global warming that, apparently, could only be ameliorated by developing markets for environmental services such as carbon sinks.

But as we have seen, Nupan was not particularly adept at the representational politics of project development and associated property claims. It was unable to control the interpretation of events associated with its project's limited authorisation by sections of the state and landowners. The mainstream media — the primary source of critical material about Nupan's project — transformed the project into a corruption narrative and undermined Nupan's viability. But this critical coverage had its own limitations. It typically abstracted carbon from the region's recent history of development and, as a result, minimised the influence of other developers competing for control over the Kamula Doso. The Kamula and other residents of the project area emerged mainly as ciphers for reified categories of understanding — they were 'indigenous', embedded in 'custom' and 'victims' of Nupan's scam. The coverage often failed to acknowledge how Kamula actively supported various investors in the often exuberantly illegal, sometimes dangerous, wealth-creating politics of development in the Kamula Doso area. Also ignored by the global media were indications of how the Kamula made sense of Nupan's carbon project by transforming elements of globally circulating carbon trading and climate change discourses into new definitions of their property interests in carbon (Filer and Wood 2012; Wood 2013).

Irrespective of the importance of these more distinctly Kamula understandings of the carbon economy, the Kamula's development options are still dependent on what the government and companies such as Rimbunan Hijau and IT&S decide to do. However, as this chapter has stressed, there is no inevitability to the realisation of any proposal concerning the future development of Kamula Doso's forests. But moves to create forms of wealth from within Kamula Doso, and privatise, or in Kamula terms, 'steal', such assets, will no doubt continue.

In 2014 VISTA Trading Ltd announced on the internet the existence of the Kamula Doso Virgin Rainforest Project. Surrounded by beautiful images of forest from more temperate regions than those found in the Kamula Doso area, the site presents Kamula Doso Block 1 as the 'Paradise Forest' of PNG, as 'pristine' rainforest, covering an area of 268,400 hectares that could be a REDD scheme. What might happen to Blocks 2 and 3 of Kamula Doso is not indicated. But we are assured 'supporting the Kamula Doso project helps conserve native rainforests and protects biodiversity on a massive scale as well as local communities'. VISTA no longer exists on the internet. This may reflect the increasingly limited life span of PNG-based carbon credit projects constituted largely as media spectacles. But despite such passing visions of the timber in the Kamula Doso concession, this timber, now perhaps freer from media distortions, is still not definable as a sustainable resource.

Acknowledgements

Along with presenting this to the Association for Social Anthropology in Oceania meeting in Washington in 2010, an even earlier version of this chapter was outlined to a seminar run by James Cook University's School of Arts and Social Sciences. Funds that helped develop this chapter were provided by the School of Arts and Social Sciences, the Australian Research Council and Carbon Planet. I owe a big debt to Wisapiye Susupiye for generously updating me on events, sometimes over the phone and sometimes in face-to-face conversations in Port Moresby. I also have the same kind of debts to many Kamula from Wawoi Falls, especially Paul Sasae, Donald Kaula and Bruce Sasai.

References

Bryan, J., P. Shearman, J. Ash and J.B. Kirkpatrick, 2010. 'Impact of Logging on Aboveground Biomass Stocks in Lowland Rain Forest, Papua New Guinea.' *Ecological Applications* 20(8): 2096–103.

Fay, D., 2013. 'Neoliberal Conservation and the Potential for Lawfare: New Legal Entities and the Political Ecology of Litigation at Dwesa–Cwebe, South Africa.' *Geoforum* 44: 170–81.

Filer, C., 2011. The New Land Grab in PNG. Paper presented to international conference on 'Global Land Grabbing', Institute of Development Studies, UK, 6–8 April.

Filer, C., R.J. Keenan, B.J. Allen and J.R. McAlpine, 2009. 'Deforestation and Forest Degradation in Papua New Guinea.' *Annals of Forest Science* 66(813).

Filer, C. and M. Wood, 2012. 'The Creation and Dissolution of Private Property in Forest Carbon: A Case Study from Papua New Guinea.' *Human Ecology* 40(5): 665–77.

Fox, J.C., C.K. Yosi, P. Nimiago, F. Oavika, J.N. Pokana, K. Lavong and R.J. Keenan, 2010. 'Assessment of Aboveground Carbon in Primary and Selectively Harvested Tropical Forest in Papua New Guinea.' *Biotropica* 42: 410–9.

Garland, E., 2008. 'The Elephant in the White Room: Confronting the Colonial Character of Wildlife Conservation in Africa.' *African Studies Review* 51(3): 504–33.

Greenpeace, 2004a. *Forest Crime File: Country Profile Update PNG.* Viewed 13 April 2015 at www.greenpeace.org/international/Global/international/planet-2/report/2004/12/forest-crime-file-rimbunan-hi.pdf.

Greenpeace, 2004b. *The Untouchables: Rimbunan Hijau's World of Forest Crime and Political Patronage.* Amsterdam: Greenpeace International.

Igoe, J., 2010. 'The Spectacle of Nature in the Global Economy of Appearances: Anthropological Engagements with the Spectacular Mediations of Transnational Conservation.' *Critique of Anthropology* 30(4): 375–97.

Igoe, J., K. Neves and D. Brockington, 2010. 'A Spectacular Eco-Tour around the Historic Block: Theorising the Convergence of Biodiversity and Capitalist Expansion.' *Antipode* 42(3): 486–512.

Independent Forestry Review Team, 2001. 'Review of Forest Harvesting Projects Being Processed towards a Timber Permit or a Timber Authority: Observations and Recommendations.'

Independent Forestry Review Team, 2003. 'Review of Wawoi-Guavi Blocks 1, 2 and 3 (Consolidated) and Vailala Block 2 & 3.'

Lane Poole, C.E., 1925. *The Forest Resources of the Territories of Papua and New Guinea.* Commonwealth of Australia: Government Printer for the State of Victoria.

Lang, C., 2010. 'REDD and Violence against Indigenous Leader in Papua New Guinea.' REDD Monitor. Viewed 23 April 2015 at www.redd-monitor. org/2010/01/15/redd-and-violence-against-indigenous-leader-in-papua-new-guinea/.

Lohmann, L., 2009. 'Regulation as Corruption in the Carbon Offset Markets: Cowboys and Choirboys United.' Viewed 13 April 2015 at www. durbanclimatejustice.org/articles/regulation-as-corruption-in-the-carbon-offset-markets-cowboys-and-choirboys-united.html.

Lowrey, K., 2008. 'Incommensurability and New Economic Strategies among Indigenous and Traditional Peoples.' *Journal of Political Ecology* 15(1): 61–74.

Macquarie Bank (Capital Advisors Limited), 2008. 'Avoided Deforestation: Kamula Doso and Western Province.' Draft manuscript. Sydney: Macquarie Group.

Melick, D., 2010. 'Credibility of REDD and Experiences from Papua New Guinea.' *Conservation Biology* 24(2): 359–61.

Mirou, N., 2013. 'Commission of Inquiry into Special Agriculture and Business Lease (C.O.I SABL) Report.' Port Moresby: Department of the Prime Minister and National Executive Council. Viewed 13 April 2015 at www.coi.gov.pg/sabl.html.

Nadarajah, T., 1993. *The Sustainability of Papua New Guinea's Forest Resource.* Boroko: National Research Institute (NRI Discussion Paper 76).

Ombudsman Commission of PNG, 2002. 'Investigation into the Decision of the National Forest Board to Award Kamula Doso FMA to Wawoi Guavi Timber Resource Permit. Final Report.'

Ombudsman Commission of PNG, 2004. 'Status Report on the Implementation of Recommendations Made in the Final Report of an Investigation into the Decision of the National Forest Board to Award Kamula Doso FMA to Wawoi Guavi Timber Resource Permit.'

PNG Forestry Review Team, 2001. 'Individual Project Review Report Number 16. Kamula Doso (Western Province).'

RH PNG (Rimbunan Hijau PNG), 2010. 'Climate Change: One-World Government Conspiracy.' *RH Group Newsletter*, Issue 16, October 2009– January 2010. Viewed 10 January 2010 at rhpng.com.pg/PDF_files/RH%20 NEWSLETTER%20ISSUE%2016.pdf.

Rose, C., 1994. *Property and Persuasion: Essays on the History, Theory and Rhetoric of Ownership*. Boulder (CO): Westview.

Shearman, P.L., J. Ash, B. Mackey, J.E. Bryan and B. Lokes, 2009. 'Forest Conversion and Degradation in Papua New Guinea 1972–2002.' *Biotropica* 41(3): 379–90.

Shearman, P., J. Bryan, J. Ash, B. Mackey and B. Lokes, 2010. 'Deforestation and Degradation in Papua New Guinea: A Response to Filer and Colleagues, 2009.' *Annals of Forest Science* 67(300).

The 2003/2004 Review Team, 2004a. 'Towards Sustainable Timber Production — A Review of Existing Logging Projects: Final Report.' (Appendix 10. Appropriateness of Current Legislative and Regulatory Requirements).

The 2003/2004 Review Team, 2004b. 'Observations Regarding the Extension of the Term of Timber Permit 1–7 Wawoi Guavi and the Proposed Alterations to the Permit Terms and Conditions.'

The 2003/2004 Review Team, 2004c. 'Attachment 1. Findings and Recommendations of the 2002/2003 Review Team Regarding TP 1–7 Wawoi Guavi — March 2003 in Observations Regarding the Extension of the Term of Timber Permit 1–7 Wawoi Guavi and the Proposed Alterations to the Permit Terms and Conditions.'

The 2003/2004 Review Team, 2004d. 'Attachment 2. General Order of Events in Observations Regarding the Extension of the Term of Timber Permit 1–7 Wawoi Guavi and the Proposed Alterations to the Permit Terms and Conditions.'

Tsing, A., 2005. *Friction: An Ethnography of Global Connection*. Princeton (NJ): Princeton University Press.

Tumu Timbers Development Limited, 2010. 'Kamula Doso Improved Forest Management Carbon Project.' Project design document for validation at Climate, Community & Biodiversity Alliance (CCBA). Viewed 14 April 2015 at www.climate-standards.org/2010/06/29/kamula-doso-improved-forest-management-carbon-project/.

Wood, M., 2009. 'A Preliminary Overview of Landowner Interests in the Kamula Doso Concession.' Unpublished report to Carbon Planet.

Wood, M., 2013. 'Spirits of the Forest, the Wind and New Wealth: Defining Some of the Possibilities, and Limits, of Kamula Possession.' In A. Aikhenvald and B. Dixon (eds), *Ownership and Possession*. Oxford: Oxford University Press.

10. 'Evergreen' and REDD+ in the Forests of Oceania

JENNIFER GABRIEL

Forests, Conversions and Climate Change

> Any global agreement on climate change will impact on how we manage our forestry operations. If we are to sustain our industry into the future, we must stay alert and aware of these developments. Most importantly, we must respond by making our case to Governments and business partners around the world. If we don't, the future of our industry will be in peril. (Tiong Hiew King 2008)

Throughout most of industrialised history, forests have been worth more felled than standing. Countries in the developed world, including Australia, New Zealand, United States, Canada and Europe have clear-felled around half of the world's primary forests to plant agricultural crops, create urban centres, extract wood for construction or burn timber as fuel. In response to the international demand for timber and timber products, most deforestation in the twentieth century has occurred in developing countries, which collectively harbour around half of the world's remaining natural forest. In the twenty-first century, predictions of global warming have led to a new way of seeing forests as repositories of carbon. In the climate change 'crisis', the maintenance of existing forests as well as increasing forest coverage is touted to make an important contribution to the mitigation of global warming, but this potential is yet to be realised (Nabuurs et al. 2007; Capoor and Ambrosi 2008). From a governance perspective, the international climate regime is playing a dominant role in influencing environmental and developmental policy internationally (Maguire 2010: 5) and is opening up spaces for critical change in 'sustainable forest management' (SFM).

To illuminate how this focus has generated new possibilities for change in Oceania's forestry industry, I first situate the idea of climate change within the immense institutional, managerial and surveillance apparatus focused on the regulation of atmospheric emissions. I then show how the 'triple-interface' (Boykoff 2010: 397) of science, policy and media is being utilised by governments, industry and corporate interests to leverage support for developmental objectives. Finally, I outline Rimbunan Hijau's responses to climate capitalism and highlight local strategies of resistance that open the company to external scrutiny.

The Idea of Climate Change

The ecological problem of industrialisation is not new, and its negative effects on nature have been debated for the past 150 years. Icons of the ecological dilemmas facing the world have ranged from deforestation in European countries in the nineteenth century, wilderness conservation at the beginning of the twentieth century and resource depletion in the 1970s (Hajer 1996: 247). Anxieties about artificially induced climate change and species extinction date back to the mid-1700s, thus the current awareness of a global environmental threat is largely a reiteration of a set of ideas that had reached full maturity over a century ago (Grove 1990: 14). What is new in the twenty-first century is the positioning of global warming as 'the greatest known challenge facing the world' (Hunt 2009: 1). As repositories of carbon, forests are seen as timely assets (Ferry and Limbert 2008) that can save the world from 'ecological apocalypse' (Ecological Internet 2011).

Climate change is an idea that has served many purposes and continues to do so (Hulme 2009). The idea of scientific mastery over climate for the global good first appeared in the closing paragraph of Ellsworth Huntington's 1915 book *Civilisation and Climate*. In reflecting the scientific ideals of the Enlightenment, Huntington wrote, 'If we can conquer climate, the whole world will become stronger and nobler' (Hulme 2009: 21). Two decades earlier, the first official explanation of how carbon dioxide can act as a greenhouse gas was provided in 1896 by the Swedish scientist Svante Arrhenius (Houghton 2004: 17). It wasn't until the mid-twentieth century that the first monitoring of increasing levels of carbon dioxide in the atmosphere began, followed by a series of studies carried out by the United States Department of Energy in the 1970s raising concerns about possible global warming (Lohmann 2006: 35). This marked the beginning of the discursive construction of global climate change.

In 1988, the United Nations General Assembly adopted resolution 45/53 on the 'Protection of global climate for present and future generations of mankind' (Lutken and Michaelowa 2008: 1), which led to the formation of the Intergovernmental Panel on Climate Change (IPCC) — the largest international scientific assessment of climate change. Although the IPCC signalled that the 'science is settled' on global warming (IPCC 2007), its critics regard the science advice of the IPCC as 'politically cast' and thereby fundamentally flawed (Carter 2010: 26). The United Nations Framework Convention on Climate Change (UNFCCC) formalised the Kyoto Protocol in 1997, which expired in 2012. The protocol's core aim was to reduce greenhouse gas emissions in the atmosphere in order to 'prevent dangerous anthropogenic interference with the climate system' (UNFCCC 2008: 3). As a regulatory mechanism, the Kyoto Protocol not only created the justification for the commodification of the atmosphere, but also inspired new forms of activism and sites of scientific, economic, social and political conflict.

'Air Money', REDD and REDD+

The Melanesian term 'air money' or 'sky money' has come to symbolise the mystique of capturing carbon dioxide (Wood 2011). Mystique is also attributed to the ways in which anthropogenic contributions to climate change have been calculated (Kellow 2007: 47), with computer models predicting future climate scenarios according to the assumptions programmed into them (Carter 2010: 21). These predictions have generated detailed economic forecasting and policy creation around the role of forests in reducing emissions from deforestation and forest degradation (REDD).

REDD first emerged in the UNFCCC negotiations over accounting rules for the impact on carbon sinks of changes in land use, particularly afforestation, reforestation, deforestation and forest management (Articles 3.3 and 3.4 of the Kyoto Protocol). In the Kyoto Protocol, a decision was made to exclude carbon emissions from tropical deforestation. Countries met their Kyoto targets primarily through national measures, but were also offered an additional means of meeting targets by way of three market-based 'emissions trading' schemes known as the carbon market, clean development mechanism (CDM) and joint implementation.

The mechanism known as joint implementation enabled Annex B parties[1] with an emission-reduction or limitation commitment to earn emission-reduction units from an emission-reduction or emission-removal project in another Annex B party, each unit equivalent to one tonne of carbon dioxide (UNFCCC 2010).[2] Credits from avoided deforestation were excluded due to the challenges and uncertainties of quantifying forest-sector emissions ('leakage'), which could potentially weaken the overall strength of the climate regime. Moreover, developing countries were concerned that a plan to reduce deforestation would threaten their sovereignty over land-use decisions and right to develop. An eligibility criterion was 'additionality', which required that reductions in emissions must be additional to any that would occur in the absence of the certified project activity (Kyoto Protocol, Article 12[5]).

The CDM allows industrialised countries to earn carbon credits from reforestation and afforestation projects in developing countries. The idea encapsulated in the CDM is that carbon markets will provide an efficient system for reducing global emissions and drive investment towards the cheapest reductions. By 2008, this mechanism had not greatly favoured forestry projects. Only one afforestation project had been approved out of the 1016 CDM projects, but approximately 100 more were in preparation (UNFCCC 2008). By 2010 there were 5,600 projects in the CDM pipeline, of which 2801 were registered and 56 others awaiting registration (UNFCCC 2011). Of these, only 14 were afforestation and reforestation projects, and another 2 were seeking registration. Only one CDM project combined carbon emissions credits with reforestation (ibid.).[3] By 2013 there were 6,838 CDM projects registered, with only 7 described as afforestation/reforestation projects.[4] By 2015 just 2 of the 7,644 registered projects were listed as afforestation/reforestation projects.[5] 'Free-trade' advocates believe that the responsibility for the effective failure of CDM in facilitating afforestation and reforestation projects lies primarily with the rules for 'additionality', which diminish financial incentives for undertaking CDM projects (World Growth 2010).

---

1   Annex B countries include the main European Union member states, Australia, Japan, Russia and the United States. Thirty-nine industrialised countries are listed in Annex B of the Kyoto Protocol.

2   As of 2010, there were 197 active joint implementation projects (UNFCCC 2010).

3   Bolivia (Project 2510).

4   India (Project 2345), China (Project 2700), Chile (Project 3338), Uruguay (Project 3845), Brazil (Project 3887), Congo (Project 4176), Uganda (Project 4653).

5   Brazil (Project 3887) and China (Project 9563); cdm.unfccc.int/index.html (viewed June 2015).

Forests only became pivotal to reducing global emissions in December 2005 when, at the United Nations Conference of the Parties in Montreal, the Coalition for Rainforest Nations,[6] led by Papua New Guinea and Costa Rica, proposed the inclusion of carbon finance for reducing national rates of deforestation. The initial proposal was limited to reducing carbon emissions from deforestation (RED) but was expanded at Bali, in December 2006, to include forest degradation (REDD) (Karousakis and Corfee-Morlot 2007; IUCN 2008; Miles and Kapos 2008). Representing the members of the Coalition for Rainforest Nations, Kevin Conrad, who was lauded as a 'hero of the environment' (Stiglitz 2008) and a 'champion of the earth',[7] tabled the idea of paying for the conservation of forests. Conrad issued a bold challenge to the United States: 'If, for some reason, you're not willing to lead, leave it to the rest of us, and please get out of the way' (Stiglitz 2008). A year earlier in January 2005, Conrad persuaded Michael Somare, then prime minister of Papua New Guinea (PNG), to call for the establishment of a coalition for rainforest nations, which he did in a speech to the World Leaders Forum at Columbia University (Filer 2011b).

In principle, REDD enables developing countries to receive financial payments from developed countries for reducing national deforestation rates below a baseline level prior to 1992. With little objection from the international community, REDD became the financial mechanism adopted at Bali to reduce emissions from deforestation and forest degradation in developing countries. The inclusion of REDD in a post-Kyoto framework was widely accepted by the international political community because it provided a way for developed countries to offset carbon emissions and enabled developing countries with large forestry resources, such as PNG, to receive payments for conservation.

Following the widespread acceptance of the principles of REDD as a carbon mitigation measure at Bali, the response of developing countries, including Africa and Malaysia, shifted towards a defensive attitude for their forestry industries. Africa demanded that REDD encompass clear recognition of the role of commercial forestry in development and conservation (World Growth 2009), while Malaysia's submission on REDD argued for a critical distinction to be made between 'deforestation' and SFM:

> Malaysia … feels that the definition of deforestation needs to be broad enough to cover the various levels and patterns of forest degradation. This is important

---

6   The Coalition for Rainforest Nations includes Bangladesh, Belize, Bolivia, Cameroon, Central African Republic, Colombia, Costa Rica, Democratic Republic of the Congo, Dominican Republic, Ecuador, El Salvador, Equatorial Guinea, Fiji, Gabon, Ghana, Guatemala, Guyana, Honduras, Indonesia, Kenya, Lesotho, Liberia, Madagascar, Malaysia, Nicaragua, Nigeria, Pakistan, Panama, Papua New Guinea, Paraguay, Peru, Republic of the Congo, Samoa, Sierra Leone, Solomon Islands, Suriname, Thailand, Uganda, Uruguay, Vanuatu and Vietnam (Wainwright et al. 2008).

7   www.unep.org/champions/laureates/2009/conrad.asp#sthash.YR9Jt1WC.dpbs (viewed June 2015).

as any level of degradation exists on the continuum between completely sound, protected forests and complete deforestation. As such, a pattern of continued forest degradation will contribute significantly to a net increase in emissions, eventually culminating in complete deforestation and should therefore be differentiated from sustainable forest management. (MNRE 2007)

This emphasis on graduated deforestation seems to imply that anything less than complete deforestation may constitute SFM; however, the International Tropical Timber Organization (ITTO) principles for sustainable tropical forest management include the wider political, social and economic criteria without which sustainability is probably unattainable. Lobbying for SFM as a mode of carbon management was pursued by global neoliberal institutions such as the ITTO, which issued a statement at Bali outlining its commitment to advancing as quickly as possible the full implementation of SFM in the tropics 'including reducing emissions from deforestation and forest degradation, carbon sequestration through restoration, thus contributing to addressing the issue of climate change'.[8] As a result of industry lobbying and political pressure from developing countries such as Malaysia, paragraph 1(b)(iii) of the Bali Action Plan included the provision for the 'sustainable management of forests and the enhancement of forest carbon stocks' (UNFCCC 2007) — this is the moment when REDD+ emerged. The critical text for the 'plus' in the Bali Action Plan, which attempted to bridge the long-standing tensions between conservation and development goals, appears after the semicolon:

> Policy approaches and positive incentives on issues relating to reducing emissions from deforestation and forest degradation in developing countries; and the role of conservation, sustainable management of forests and enhancement of forest carbon stocks in developing countries (UNFCCC 2007).

The inclusion of forest conservation in an international voluntary agreement to address the impacts of global warming was driven by estimates from British economist Lord Stern, who attributed more than 18 per cent of global carbon emissions to deforestation (Stern 2006: xxv). According to free-trade lobbyist Alan Oxley of World Growth, independent research by United States-based consultants Winrock International (commissioned by the World Bank) estimated that deforestation accounts for only 6–8 per cent of global emissions (World Growth 2009). Broadening the definition of 'forests' to include plantation forests, the Food and Agriculture Organization of the United Nation's *State of the World's Forests 2011* (FAO 2011: 3) claimed that globally, the overall rate of deforestation is slowing,[9] partly due to a transition from deforestation to

---

8   Statement by Emmanuel Ze Meka, Executive Director of the ITTO at the High-Level Segment of the Thirteenth Conference of the Parties to the UNFCCC – Cop 13/CMP 3, Bali, Indonesia, 3–14 December.

9   The FAO found that global deforestation and loss of forests from natural causes ('while still alarmingly high') has 'decreased from an estimated 16 million hectares per year in the 1990s to around 13 million hectares per year in the last decade' (FAO 2011: 3).

afforestation in the Asia–Pacific region, where plantation forest area increased by 2.85 per cent annually over the past decade (ibid.: 9).[10] According to the FAO this was accompanied by an increase of around 1.92 per cent (from 2000 to 2010) in the global area of forest designated primarily for the conservation of biological diversity (ibid.: 10).[11]

Climate Capitalism and Industry Lobbying: From Bali to Copenhagen and Beyond

As the Conference of Parties travelled from Bali (2007) to Copenhagen (2009), the UNFCCC attempted to garner consensus for the replacement to the Kyoto Protocol. Negotiators drafted and redrafted the policy text for REDD, determined to produce a coherent vision for the international community and generate a set of norms and practices that would govern national carbon emissions. However, these forums also afforded opportunity for free-trade advocates and industry lobby groups to present arguments in support of the role of industrial forestry and plantation development to be included in global warming solutions. Powerful industry organisations leveraged on the FAO definition of 'forests', arguing that industrial plantations should qualify as forest conservation projects as long as they do not contribute to a net increase in national carbon emissions. Consensus-building for industrial plantations was a strategy of the global timber industry. Among the prominent and influential advocates for the industrial forestry sector was Alan Oxley, managing director of international trade consultants ITS Global, an Australian public relations company. ITS Global has been commissioned by Rimbunan Hijau since 2006 to counter the ongoing campaign by Greenpeace (2004, 2006) and other non-governmental organisations (NGOs) against industrial logging in PNG. Oxley is a former diplomat who held a position on Australia's Foreign Affairs Council as Australia's envoy to the General Agreement on Tariffs and Trade, the predecessor to the World Trade Organization in the late 1980s.[12] Oxley describes himself as 'one of the world's leading experts on globalization and international trade' and is the founding chairman of 'World Growth', a free-market NGO based

---

10 'This was primarily due to large-scale afforestation efforts in China, where the forest area increased by 2 million hectares per year in the 1990s and by an average of 3 million hectares per year since 2000. Bhutan, India, the Philippines and Viet Nam also registered forest area increases in the last decade' (FAO 2011: 9).

11 The FAO claims that the area of forest designated primarily for conservation of biological diversity in Asia and the Pacific increased by 1.8 per cent from 2000 to 2010 (FAO 2011: 10, Table 9).

12 Alan Oxley is currently chairman of the Asia–Pacific Economic Cooperation (APEC) Study Centre at the Royal Melbourne Institute of Technology University, Australia.

in Washington DC, producing targeted industry research, particularly in areas such as Malaysia and PNG, with a direct bearing on international trade issues linked to oil palm and timber production.

World Growth uses high-impact international business media to disseminate pro-industry 'solutions' to leaders of developed and developing countries, and draws upon moral arguments about progress and growth to justify its lobbying:

> World Growth believes that helping the developing world realize its full potential is one of the great moral aims for those of us fortunate to live in the wealthy developed world (www.worldgrowth.org).

Oxley's use of multimedia to mobilise contested scientific claims in support of plantation forestry and the need for adaptation through further deforestation ('alternative land use') relies largely on the language of uncertainty and risk management. During the international climate negotiations, his outspoken views on pro-development solutions to global warming reached international business audiences and government leaders via Bloomberg, CNN, CNBC, *The Wall Street Journal*, *International Herald Tribune*, SBS World News Australia, *Jakarta Post*, *Business Times* and *The New York Times*, as well as high-impact trade journals. The intense international media campaign mounted by World Growth attempted to undermine the fundamental principles of REDD as a conservation strategy and questioned the ethics and motives of some of the largest environmental lobby groups including Greenpeace International and WWF (World Wide Fund for Nature). Although World Growth's campaign against the green movement relentlessly targeted the agendas of international environmental organisations,[13] Oxley's critique of institutions that limited growth in developing countries also extended to the World Bank (World Growth 2011b) and the United Nations:

> We are frequently opposed by other non-governmental organizations (NGOs) who are ideologically unreceptive to free enterprise and skeptical of the merits of economic growth. Along with NGOs we face challenges at major multilateral institutions such as the United Nations which too often promote one-size-fits-all policy responses to problems. (Chairman's Message, worldgrowth.org)

In October 2010, Oxley's pro-development stance drew strong criticism from eminent members of the scientific community. In 'An Open Letter about Scientific Credibility and the Conservation of Tropical Forests', Laurance et al. (2010) alleged that ITS Global and World Growth 'have at times treaded a thin line between reality and a significant distortion of facts' (ibid.: 1). Particular criticism was directed at Oxley's lobbying support for Rimbunan Hijau, which was described as a 'predatory corporation' (ibid.). Oxley attacked the article's credibility, claiming it 'simply re-counts — almost source for source

---

13   See worldgrowth.org/category/research/ (viewed June 2015).

— Greenpeace's failed campaign to shut down the PNG forest industry from last decade' (ITS Global 2011a). Oxley staunchly defended the PNG forestry industry and his client, Rimbunan Hijau:

> Laurence et al. throw a litany of accusations against PNG's forest industry — specifically one company, Rimbunan Hijau — and align it with the country's myriad social and environmental problems: corruption, violence, poor development indicators, deforestation and weak governance. The implication is that without the forest industry in Papua New Guinea, the country would somehow magically be free of corruption, deforestation and other ills. Anyone who has set foot in Papua New Guinea would be acutely aware of the absurdity of this. (ITS Global 2011a)

Around this time, PNG was emerging from a 'carbon cowboy' fiasco (see Wood, Chapter 9, this volume), which had caused a media sensation two years earlier and brought the PNG government into disrepute at the much anticipated Copenhagen Climate Talks. Leading up to the UNFCCC conference at Copenhagen (COP15), touted as a crucial event in the negotiating process, dichotomising discourses of 'conservation' and 'sustainable development' had generated serious conflicts between countries of the North and the South. By the time world leaders met in Denmark in 2009, scientific uncertainty and conflicting value claims had created irreconcilable differences in the positions adopted by developed and developing countries. The bitterness and scepticism that bedevilled the talks became exasperated by a leaked draft agreement from the European Union, which called for a total halt to deforestation by 2030. This evoked further anger from the tropical countries at the North's failure to commit to dramatic emissions reductions. The subsequent result in the Copenhagen Accord was the removal of all targets for deforestation and a much weakened language on safeguards. An earlier draft contained a target for reducing deforestation by at least 50 per cent by 2020, but as a result of the leaked European Union proposal, deforestation targets were removed from the negotiating text (Lang 2009). The accord firmly endorsed REDD+ with an emphasis on SFM consistent with long-term sustainable land management (Verchot and Petkova 2009). Governments of the world's biggest palm oil producers, including Malaysia, swiftly rejected proposals put forward by global environmental groups to exclude the expansion of oil palm plantations in REDD+ projects.

A year later, in December 2010 at the climate change negotiations at Cancún in Mexico, the UNFCCC further endorsed the option for developing countries to conserve their forests and receive payments for avoided deforestation and the enhancement of carbon stocks (REDD+). Emission credits could also be granted for the use of 'green technology' developed through the CDM. In addition to having no binding commitments to curb deforestation, the inclusion of plantation forests as an internationally endorsed modality of carbon capture and storage was one of the most controversial inclusions in the Cancún Agreement. In COP 16, Cancún 2010, the final decision on REDD+:

> Encourages developing country Parties to contribute to mitigation actions in the forest sector by undertaking the following activities, as deemed appropriate by each Party and in accordance with their respective capabilities and national circumstances:
>
> (a) Reducing emissions from deforestation;
>
> (b) Reducing emissions from forest degradation;
>
> (c) Conservation of forest carbon stocks;
>
> (d) Sustainable management of forest;
>
> (e) Enhancement of forest carbon stocks.[14]

The agreement emphasised the need to ensure at national and community levels that REDD+ 'complements restoration, poverty alleviation and adaptation' and 'promotes and operationalizes safeguards and accountability' (IISD 2011). These priorities couple the ideals of SFM with the free-market principles of ecological modernisation.

REDD and Ecological Modernisation

In 2011, as developing countries began committing voluntary targets to reduce greenhouse gases and participate in international REDD programs to cease deforestation and reshape their economies as 'low carbon' economies, Alan Oxley released a study by World Growth which argued that conservation strategies such as REDD will impoverish the poor and put biodiversity at risk (World Growth 2011a).

The sentiments expressed by Oxley are closely echoed in Rimbunan Hijau's (PNG) pro-development discourse, which draws upon principles of ecological modernisation to champion economic growth and global trade over conservation strategies:

> The economic consequences of global warming mitigation strategies currently proposed will probably be worse than the effects of global warming itself. Therefore, adaptation and resiliency strategies should be considered as a more cost-effective alternative. In addition, 'no regrets' strategies that will provide benefits from greater economic growth whether global warming proves to be a problem or not, should be adopted at once. (RH PNG 2010: 4)

---

14 Report of the Conference of the Parties on its sixteenth session, held in Cancún from 29 November to 10 December 2010, FCCC/CP/2010/7/Add.1. (unfccc.int/resource/docs/2010/cop16/eng/07a01.pdf#page=2).

Ecological modernisation theorists argue that forest destruction and biodiversity loss can be reversed by developing flexible liberal institutions which foster sustainable capitalism through the restructuring of the state towards participatory policy making. Rather than a withering of the role of the state in environmental governance, the introduction of favourable conditions for collaboration with civil society, environmental groups and the public sphere is expected to transform responsibilities, incentives and roles to the market, reorienting the state to serve as a coordinating function with other sectors in support of forests (Mol 1995: 46–7). Ecological modernisation, which has dominated or been deployed by industry advocates in political debates concerning global ecological objectives (Hajer 1996: 249), strives to achieve a balance between environmental sustainability and economic development. Ecological modernisation is increasingly articulated in the principles of SFM. For example, in the past decade the discourse of SFM has gradually shifted from 'forest management by exclusion' to 'management by inclusion of stakeholder groups', and 'sustained yield timber management' has become 'sustainable forest ecosystem management' (Kant and Lee 2004). Such ecosystem-based terminology (some would call rhetoric) is underpinned by concepts such as 'system', 'relationships', 'interconnectivity', 'community', 'uncertainty', 'adaption', 'precaution' and 'inclusivity', which now permeate natural resource management discourses (Gale and Haward 2011: 46). In the international regulation of SFM, Maguire (2010: 288) argues that the design of REDD+ 'is the first global instrument to embrace the ideas behind environmental justice in the forest context'.

According to Kant and Lee (2004: 215), there are two dominant features of SFM under a REDD framework which offer potential for moving towards a more sustainable future in Oceania's forests and local communities. First, REDD can include the recognition of multiple forest values, beyond the customary timber values, including non-timber products, value of eco-services and indigenous values (ibid.). Second, SFM under REDD+ may accommodate the preferences of multiple stakeholder groups, such as local communities and environmental groups in forest management decision making (ibid.). The problem of ecological modernisation–influenced SFM development under the framework of REDD+ is that it is designed to be implemented in a classical top-down approach, starting at government level and ending with the integration of rural communities living in or around the forests. Thus, stakeholder participation is the last consideration in project implementation, and although safeguards are being negotiated to ensure that forest actors have a reasonable share in the economic benefits, the capacity to enforce compliance depends on the SFM system established at the national level. This means that SFM implementation largely relies on voluntary corporate commitments to 'best practice' standards in natural resource management. In climate capitalism, the rhetoric of SFM has acquired authority and served particular social interests, and is increasingly becoming institutionalised within organisational structures and practices that attempt to legitimise the expansion of corporate capitalism.

Rimbunan Hijau: REDDish Transformations in Green Capitalism

In its home state of Sarawak, Malaysia, Rimbunan Sawit Berhad (a publicly listed company majority owned by the Rimbunan Hijau Group) is one of the leading palm oil producers and processors of crude palm oil and oil palm seeds, with approximately 93,500 hectares of land in Miri, Kuching and Sibu (Sarawak).[15] 'Sustaining Wellness' is the slogan on its annual reports for 2013 and 2014 (RSB 2013, 2014). Adopting a policy of 'Developing Smart Partnerships' (RSB 2009: 9), Tiong Hiew King, (then) executive chairman, focused on equity acquisitions in integrated oil palm plantations in Sarawak. Rimbunan Sawit's 'group vision' is to be one of the leading integrated oil palm companies in the Asia–Pacific. Much of the demand for palm oil is being generated by the global demand for alternative energy sources, such as biofuel. In the 2008/09 financial year the group increased its land area by 53.86 per cent (RSB 2009). This was followed in the 2009/10 financial year with equity acquisition of 11 plantation companies, and a plantation estate with commercial rights.

In tandem with its 'group mission' to enhance stakeholder value, Rimbunan Sawit deploys the language of climate-friendly development to assert its commitment to nature:

> One of our commitments to the nature is the Clean Development Mechanism (CDM) project in RH Plantation Palm Oil Mill. The project will recover methane caused by the decay of biogenic matter in the effluent stream. Methane recovered will be used to generate electricity for the mill and vicinity plantations. The project will reduce greenhouse gas emissions in an economically sustainable manner and is expected to reduce carbon dioxide equivalent by 20,002 tons per year. Zero burning during land development is practised to ensure minimal carbon emission. Besides, biological control is practised for pest management to reduce the dependency on pesticides. Empty fruit bunch is used as much [mulch] in plantations in order to recycle the nutrients present. This helps to reduce the dependency on chemical fertilizers. (RSB 2009: 10–1)

15　In 2013, the production area was 36,867 hectares, an increase of 1,739 hectares from 2012. The group's oil palm–planted area was 54,659 hectares (RSB 2013). In 2014, the group's oil palm–planted area increased to 57,182 hectares (RSB 2014).

Another of Tiong Hiew King's majority-owned public companies in Malaysia, Subur Tiasa Holdings, says it has championed 'eco-friendly' practices and adopted SFM because of its 'love of forests and wood':

> We love forests. We love wood.
>
> Our heart for habitat compels us to commit not just to practice Sustainable Forest Management but to spearhead a comprehensive strategy that strives to ensure the sustainability of the environment. (suburtiasa.com.my/csr/the-wood-spa/)

Claims to SFM and commitment to tree planting (for sustainable supply) are positioned in Subur Tiasa's investment discourse as reflecting the ethical orientation of its director Tiong Ing, who is described as a 'zealous lover of the habitat':

> Simply put, we plant trees. We plant significantly more trees than we log. Thirty times to be precise (as of 2010). It is unrealistic to ignore the continual demand for wood as a material for very practical needs. Extensive research is also being carried out to ensure proper conservation of our forests.
>
> Selected indigenous species are being planted and rapid-growing exotic species are planted in areas designated for Industrial Tree Planting. Island Corridor Planting is also practised to reduce the burden bore by the environment and to preserve biodiversity which is at Subur Tiasa's heart.
>
> To accelerate and ensure our sustainable growth in today's competitive business environment, we have partnered with our major shareholder, Rimbunan Hijau Group, to invest in our wood treatment strategy that is fronted by none other than our Managing Director, Dato' Tiong Ing, a zealous lover of the habitat. (suburtiasa.com.my/csr/the-wood-spa/)

In New Zealand, Rimbunan Hijau's sister company Ernslaw One[16] (the fourth-largest forestry plantation owner in New Zealand) promotes its investments as being carbon neutral:

> Ernslaw One has over 25,000 ha of post-1989 Kyoto compliant forests, making it one of the largest owners of post-1989 forests in New Zealand … The forests are spread throughout New Zealand, and comprise two species, Radiata pine and Douglas fir … Radiata pine is clearfelled anytime from age 25 onwards, whereas Douglas fir is from 45 years. This means that when Ernslaw One starts harvesting the Radiata pine in approx 2022, the Douglas fir will still be growing and will cover any carbon liability that may occur. Similarly, when the Douglas fir is being harvested, the Radiata pine will be growing. Ernslaw One can thus manage their forests for timber production, as well as ensuring that no liability arises for any carbon that is sold from the forest. (ernslaw.co.nz/carbon-sequestration/)

---

16   Ernslaw One's major shareholder is the Tiong family, headed by Tiong Hiew King.

Drawing on plantation experience in New Zealand as well as in Malaysia, Rimbunan Hijau is moving towards investing in plantations in PNG. The company has been testing plantation species for suitability in Western Province (RH PNG 2009). In the proposal subtitled 'An Initiative by Rimbunan Hijau (PNG) Group Demonstrating Its Continued Commitment to Sustainable Forest Management in Papua New Guinea through Reforestation' (RH PNG n.d.), the matrix of benefits specified carbon sequestration as one of the ecological benefits (Table 10.1).

Table 10.1 Rimbunan Hijau's matrix of benefits from plantation forestry.

Social	Ecological	Economic
Employment	Carbon sequestration	Timber production
Recreation	Wildlife	Carbon trade
Taungya system	Aesthetic	Diversification of local economy
Fuelwood	Landscape and biodiversity restoration	Recreation, tourism, harvesting and marketing of minor forest products etc.

Source: RH PNG (n.d.: 19).

Deploying the language of climate-compatible development, Rimbunan Hijau (PNG) promotes the creation of timber plantations in PNG as effective carbon sinks ('enhanced stocks'):

> Plantation appears to be a good source of 'carbon sink'… As Laarman and Sedjo (1992) rightly stated, 'just as climate affects forest, forests are able to affect climate — if deforestation puts carbon dioxide in the atmosphere, reforestation takes it out'. McLaren further showed that 1-ha of radiata pine plantation in New Zealand absorbs 24 tons of carbon annually or a total of 24 million tons of carbon annually from New Zealand forest alone. Similarly, Houghton and Skole (1990) estimated that 100–200 million hectares of developing or growing forests would absorb 1 billion tons of carbon. This further indicates that a growing forest plantation will absorb more CO_2 for its growth, thus reducing the level of CO_2 in the air. (RH PNG n.d.: 19)

Through plantation production, Rimbunan Hijau sees itself as making a significant contribution to PNG's carbon mitigation efforts:

> The Company as part of its Corporate Environmental responsibility will commission an independent 3rd Party to Study Silvicultural Investments in Carbon Accounting/Abatement and further through its Reforestation & Forest Management practices will calculate forest carbon contributions to the regional & global carbon cycle. This however will also benefit Papua New Guinea in its efforts for carbon trading because in order to buy or sell a good or service, it needs to be quantified. And to quantify, we need to understand how to measure forest ecosystem carbon dynamic & carbon budgets which are certainly essential to the development of carbon markets and the company is willing to undertake this as a pilot project. (RH PNG n.d.: 19)

Rimbunan Hijau's expressions of 'concern' for the global environment, and its social commitments to the nation, have failed to win the support of non-profit civil society organisations, including PNG's Eco-Forestry Forum, which has consistently adopted an adversarial stance against the company. In 2009, the Eco-Forestry Forum rejected Rimbunan Hijau's offer to sponsor 500 mangrove trees for World Environment Day because of a court case brought about by the forum which (successfully) challenged the legal extension of Rimbunan Hijau's logging concession 'Kamula Doso' in Western Province. Rimbunan Hijau's PNG public relations manager opined: 'So much for battling the global climate change and environment sustainability war' (RH PNG 2010: 3).

As Benson and Kirsch (2010: 45–6) have cautioned, strategies used by corporations to manage or neutralise critique, including idioms of ethics, health, environmentalism and corporate responsibility, generally promote business as usual. It is also the case, however, that when corporations claim to be 'ethical' and 'sustainable' it opens up spaces for public scrutiny and critique:

> RH is a significant contributor to the nation's economic and social wellbeing. RH works closely with landowners, communities, NGOs and government agencies as part of its operations … RH also takes the environment seriously. RH is Papua New Guinea's industry leader on environmentally responsible and 100 per cent legal management of forests … RH has prided itself on an economically, environmentally and socially sustainable future for Papua New Guinea. (rhpng.com.pg)

Rimbunan Hijau's claims of corporate social responsibility have not yet resulted in a 'politics of resignation' (Benson and Kirsch 2010), and the company's attempts to legitimise its corporate power within PNG have largely failed, but Rimbunan Hijau has continued its attempt to counter negative representations through a reliance on orthodox economic understandings of 'development', 'growth' and 'global equity' (Gabriel and Wood, forthcoming). A Foucauldian critique (Lattas 2011) would suggest that the company's use of discourses, linked to SFM, corporate social responsibility and ecological modernisation, involves the pragmatic deployment of knowledge for the primary goal of governing subjects (landowner groups) and their subjectivities.

In terms of contributing to climate change solutions, Rimbunan Hijau lobbyist Alan Oxley promotes palm oil plantations as more efficient carbon sinks than natural forests:

> Palm oil is also very greenhouse friendly. Properly managed, plantations absorb more carbon dioxide than natural forest. (Oxley 2009)

To pursue oil palm development in PNG, Gilford Limited, a Rimbunan Hijau subsidiary, is seeking to secure long-term tenure property rights in East New Britain through a highly controversial Special Agricultural and Business

Lease (SABL) that bypasses forestry laws and enables deforestation through the clear-felling of primary forests. The granting of SABLs allows customary land (including primary forests) to be legally converted into agricultural development on land which is leased from landowners for up to 99 years. For this reason, SABLs have been described as a mechanism for modern 'land grabs':

> It is a moot point whether the companies interested in the acquisition of such land in PNG have any genuine interest in its agricultural potential, or whether they are simply looking for new ways to log PNG's native forests (Filer 2011a: 2).

The extent of the current land grab is staggering, involving almost 5 million hectares of customary land (11 per cent of PNG's total land area) (Filer 2011a: 2). In 2010, the Australian Centre for International Agricultural Research raised concerns that SABLs were purely a means of accessing saleable timber resources in the guise of agricultural (oil palm) development: 'there is little evidence that these proposals will lead to viable palm oil production' (Nelson et al. 2010: 11). In 2013, an academic study concluded that a large-scale land grab was occurring in PNG under the guise of oil palm (Nelson et al. 2014).

Through the SABL mechanism, Rimbunan Hijau is establishing an 'integrated rural development project' in East New Britain Province, involving timber and palm oil plantations. Ivan Lu, executive director of Rimbunan Hijau's PNG operations, claims that the rural development project that will lead its expansion into the palm oil sector in PNG will contribute some K800 million (US$300 million) in royalties, payments, levies and other community funding, as well as providing transport and social infrastructure to local communities (ITS Global 2011b). Although Rimbunan Hijau asserts that the majority of landowners support the land reform (Anon. 2011), other landowners fear the loss of control of land for more than three generations, which they believe could have devastating consequences (Chandler 2011). Some landowners assert that the Pomio SABL was fraudulently obtained and that many of the names purporting to approve the lease on behalf of landowners were local children (including a 3-year-old). Other names were said to belong to deceased villagers (*Post-Courier* 2011).

In October 2011, the Pomio SABL in East New Britain became the centre of international condemnation following allegations of police brutality and corporate complicity. Disgruntled landowners from Pomio filed a lawsuit with the Centre for Environmental Law and Community Rights (CELCOR) against Rimbunan Hijau and the state in the proposed Sigite Mukus Integrated Rural Development Project, accusing them of fraudulently depriving landowners of their customary land for a logging project. One of the plaintiffs, Paul Pavol, who had been leading the protest, claimed that many locals were angry because the leases were granted without the approval of the majority of traditional landowners:

> We have never given consent, we did not say yes, we want oil palm, and we did not even sign any documents. I think we are going to lose all the land [44,000 hectares] for 99 years, that's three generations. (ABC Radio Australia 2011)

The police were flown into Pomio on Rimbunan Hijau's commercial aircraft Tropic Air (between 3 and 6 October 2011) in alleged retaliation, and a number of people were reportedly intimidated and beaten with fan belts and sticks, and police locked young men in shipping containers (*Post-Courier* 2011). Pictures of placards were distributed anonymously via email condemning Rimbunan Hijau's presence in Pomio (John 2011).

In its formal response, Rimbunan Hijau defended the integrity of the State and the independence of the PNG constabulary:

> The majority of the land owners support the palm oil projects. The recent attempts by this few disgruntled land owners with assistance from NGOs, whom do not support any form of development to derail and stop this very important project, was refused by the National Court recently. The royal constabulary (of PNG) is very independent and high integrity institution of the state of Papua New Guinea and we welcome the comments of acting police commissioner Tom Kulunga.

> We regret this constant unverified adverse media release by (a) non-government organization hell bent (on) generating adverse publicity in order to keep their anti-government campaign alive for domestic and international donors. It is very much regretted and it is not helpful for a country that is seeking to develop its resources sustainably which is much needed. (Rheeney 2011)

The commission of inquiry into the granting of SABLs found that in the Pomio SABL, no legal representation of the landowners was provided during the signing of the sublease agreements, and executives of Gilford Limited appeared to have given the sublease documents to the executives of the landowner companies without any proper explanation or advice of its contents: 'RH is no small player in the forestry business. You treated this very lightly', the commissioner told the company representatives. He added he has not seen equity for the landowners: 'Unfortunately, this is not reflected in all the agreements we have seen' (Tiden 2011).

Close scrutiny of the current position of the palm oil industry in PNG further casts Rimbunan Hijau as a rogue player if it uses the SABL mechanism for oil palm development without adhering to the principles of the Roundtable on Sustainable Palm Oil (RSPO). As Filer (2011a: 22) has pointed out, the two companies responsible for PNG's entire palm oil exports (New Britain Palm Oil Ltd and Hargy Oil Palms Ltd) comply with the standards of the RSPO, pledging that their actions, directly or indirectly, do not contribute to the clearance of native forests, especially in high-conservation areas. Therefore, if Rimbunan Hijau chooses to continue its plans to develop oil palm plantations in PNG through the SABL mechanism, it needs to decide whether it supports corporate social responsibility standards set out by the RSPO and embraced by PNG's largest palm oil exporters, or whether it will pursue the agenda of its lobbyists and the PNGFIA (PNG Forest Industries Association) in supporting forest clearance for palm oil production.

Clear-felling of forests for agriculture is also at odds with the draft PNG Climate Compatible Development Policy 2013–15, which aims to 'reduce deforestation and forest degradation in commercial agriculture and forestry activities and promote afforestation/reforestation on marginal lands' (OCCD 2013).

Under the ITTO definition, Rimbunan Hijau cannot claim to be practising SFM while adopting an adversarial position in dealing with sections of the community, PNG's established oil palm industry (which complies with RSPO guidelines), as well as the international community which opposes SABLs because they are seen to be outside best practice. Although Rimbunan Hijau (PNG) says it is genuinely interested in improving corporate practice by introducing incremental reform through green technology, verification schemes and community engagement, it is yet to develop collaborative relationships with environmental or social justice NGOs. In this sense the company has adopted a largely defensive position with broader civil society.

Concluding Remarks

Despite Rimbunan Hijau's claims of social and environmental responsibility, including a commitment towards a greener future, the company's involvement in PNG's highly controversial land-use scheme (SABL) has exacerbated the perception that it is still engaging in unethical commodity production.

In May 2015, four years after the policy board of the United Nations Collaborative Programme on Reducing Emissions from Deforestation and Forest Degradation in Developing Countries (UN-REDD) approved the National Programme Document which set out how the PNG government proposed to achieve a state of 'REDD plus readiness' within three years (Filer 2011b), forestry minister Douglas Tomuriesa announced that the REDD+ mechanism in Papua New Guinea is 'ready for business'. He added that PNG is prepared to utilise the forests for REDD programs, which includes the conservation of land for REDD credits (Dateline Pacific 2015).

Signalling a move towards SFM, the National Executive Council has endorsed a total ban on log exports by 2020 (ibid.). However, PNG's policy and laws regarding REDD processes must also address fundamental issues of land tenure, ownership, prior informed consent and equitable benefit sharing. Corporate investment will need to complement domestic efforts and contribute towards enhancing SFM and sustainable development by becoming more transparent, more equitable and more accountable to local resource users.

To adhere to the principles of SFM embedded in REDD+, Rimbunan Hijau will need to work in tandem with approaches that seek to achieve global ecological sustainability and support a low-carbon economy nationally, while avoiding measures of social control that alienate sections of the community and broader civil society. This is line with Strategy 1 (7.4.2) of the draft PNG Climate Compatible Development Policy 2013–15:

> Encourage investments in natural capital such as agriculture, forestry, fisheries, water and mineral resources that promotes sustainable development and respect for community rights and livelihood whilst safeguarding national interest (OCCD 2013: 36).

A key idea of SFM under ecological modernisation theory is that markets for 'ethical commodities' can improve the ability of the private sector to internalise the environmental costs of production. Through branding and market differentiation, ethical commodities have the potential to produce economic rent (Guthman 2002, 2004), which provides a means for compensating for best practice in social and environmental management, while redistributing profit to rural communities. In terms of SFM, the success of multifunctional integrated approaches to climate capitalism such as REDD+ relies on establishing cooperative partnerships between landowners, governments, companies and non-profit civil society organisations — these are all 'brand-building' exercises that can redefine forestry values and impact corporate brands positively. Although opportunities for reform are made possible through such schemes, it is questionable whether they will bring about productive change in corporate approaches to social and environmental relationships in host countries such as PNG. As multinational companies such as Rimbunan Hijau compete for market share in the newly emerging international governance regime for forests, evidence of voluntary compliance with socially and environmentally sustainable development is increasingly necessary to demonstrate global standards of corporate social responsibility which require business activities to be compatible with robust social, environmental and ecological safeguards.

The persistence of international critique, social and environmental campaigns, and landowner resistance groups indicates that the ongoing presence of Rimbunan Hijau in the forests of Oceania will increasingly be defined by its ability and willingness to deliver outcomes defined by local perspectives and desires, as well as national and global objectives. Instead of deploying ecological-modernist discourse as a strategy to dilute the political impulse for environmental reforms (Hajer 1995), a commitment to thinking outside the borders of ecological modernisation to provide tangible and equitable benefits will need to be part of Rimbunan Hijau's strategy in order for it to deliver on its promises of corporate social responsibility and SFM. This broader strategy will need to include addressing the power imbalances between industry and landowners.

I have argued that in the expanding frontiers of global climate capitalism, new models of corporate forestry management in PNG are being challenged by the politics of 'friction' (Tsing 2005), which include ongoing forms of participatory politics. An increased focus on the use of corporate and industry discourse to influence the policy networks of Oceania's forestry industry may contribute to new opportunities for holding corporations accountable to their promises.

Acknowledgements

The author wishes to thank the Department of Anthropology at James Cook University for supporting this PhD research, and Michael Wood and Colin Filer for their thoughtful comments on an earlier version of this chapter.

References

ABC Radio Australia, 2011. 'Logging Firm Paid to Quell Violence, Say PNG Police.' ABC Radio Australia, 11 October. Viewed 11 October 2011 at radioaustralia. net.au/international/2011-10-11/416112.

Anon., 2011. 'Project Will Churn Out K800 Million for Pomio, Says RH.' *The National*, 26 October.

Benson, P. and S. Kirsch, 2010. 'Capitalism and the Politics of Resignation.' *Current Anthropology* 51(4): 459–86.

Boykoff, M.T., 2010. 'Carbonundrums: The Role of the Media.' In S.H. Schneider, A. Rosencranz, M.D. Mastrandea and K. Nuntz-Duriseti (eds), *Climate Change Science and Policy*. London: Island Press.

Capoor, K. and P. Ambrosi, 2008. *State and Trends of the Carbon Market 2008*. Washington (DC): The World Bank.

Carter, R.M., 2010. *Climate: The Counter Consensus*. London: Stacey International.

Chandler, J., 2011. 'PNG's Great Land Grab Sparks Fightback by Traditional Owners.' *The Age*, 14 October. Viewed 14 October 2011 at theage.com.au/national/pngs-great-land-grab-sparks-fightback-by-traditional-owners-20111013-1ln1m.html.

Dateline Pacific, 2015. 'PNG Putting Its Hand Up on REDD, Says Forestry Ministry.' Dateline Pacific, 28 May. Viewed 11 June 2015 at www.radionz.co.nz/international/programmes/datelinepacific/20150528.

Ecological Internet, 2011. 'Known Culprits Making Earth Uninhabitable.' E-Newsletter, 25 May. Viewed May 2011 at www.facebook.com/note.php?note_id=10150194408541242.

FAO (Food and Agriculture Organization of the United Nations), 2011. *State of the World's Forests 2011*. Rome: FAO. Viewed 10 April 2011 at fao.org/forestry/sofo/en/.

Ferry, E.E. and M.E. Limbert (eds), 2008. *Timely Assets: The Politics of Resources and Their Temporalities*. Santa Fe (NM): School for Advanced Research Press.

Filer, C., 2011a. 'The New Land Grab in PNG.' Paper presented at international conference on 'Global Land Grabbing', Institute of Development Studies, UK, 6–8 April.

Filer, C., 2011b. 'REDD-plus at the Crossroads in Papua New Guinea.' East Asia Forum, 23 July. Viewed 12 September 2011 at eastasiaforum.org/2011/07/23/redd-plus-at-the-crossroads-in-papua-new-guinea/.

Gabriel, J. and M. Wood, forthcoming. 'The Rimbunan Hijau Group in the Forests of Papua New Guinea.' *Journal of Pacific History*.

Gale, F. and M. Haward, 2011. *Global Commodity Governance: State Responses to Sustainable Forest and Fisheries Certification*. London: Palgrave MacMillan.

Greenpeace International, 2004. 'Forest Crime File: Rimbunan Hijau'. 8 December. Viewed 10 December 2008 at greenpeace.org/international/en/publications/reports/forest-crime-file-rimbunan-hi/.

Greenpeace International, 2006. 'Rimbunan Hijau Group: Thirty Years of Forest Plunder.' 30 May. Viewed 2 February 2008 at greenpeace.org/international/en/publications/reports/RH-30years-forest-plunder/.

Grove, R. 1990. 'Commentary. The Origins of Environmentalism.' *Nature* 345: 11–14.

Guthman, J., 2002. 'Commodified Meanings, Meaningful Commodities: Rethinking Production-Consumption Links through the Organic System of Provision.' *Sociologia Ruralis* 42(4): 295–311.

Guthman, J., 2004. 'Back to the Land. The Paradox of Organic Food Standards.' *Environment and Planning* 36(3): 511–28.

Hajer, M.A., 1995. *The Politics of Environmental Discourse: Ecological Modernization and the Policy Process.* Oxford: Oxford University Press.

Hajer, M.A., 1996. 'Ecological Modernisation as Cultural Politics.' In S. Lash, B. Szerszynski and B. Wynne (eds), *Risk, Environment and Modernity.* London: Sage Publications.

Houghton, J.T., 2004. *Global Warming: The Complete Briefing.* Cambridge: Cambridge University Press (3rd edition).

Hulme, M., 2009. *Why We Disagree about Climate Change: Understanding Controversy, Inaction and Opportunity.* Cambridge: Cambridge University Press.

Hunt, C.A.G., 2009. *Carbon Sinks and Climate Change: Forests in the Fight against Global Warming.* Cheltenham (UK): Edward Elgar.

IISD (International Institute for Sustainable Development), 2011. 'A Summary Report of the Ninth Rights and Resources Initiative (RRI) Dialogue on Forests, Governance and Climate Change.' *Rights and Resources Initiative Dialogue Bulletin* 173(3): 9 February. Viewed 15 June 2015 at www.iisd.ca/download/pdf/sd/ymbvol173num3e.pdf.

IPCC (Intergovernmental Panel on Climate Change), 2007. '1.2 The Nature of Earth Science. Climate Change 2007: Working Group I: The Physical Science Basis.' In *IPCC Fourth Assessment Report: Climate Change 2007.* Viewed 11 June 2015 at www.ipcc.ch/publications_and_data/ar4/wg1/en/ch1s1-2.html.

ITS Global, 2011a. 'Greenpeace's Anti-Forestry Campaign Re-Emerges at James Cook University.' Forestry & Development, 8 February. Viewed 10 February 2011 at forestryanddevelopment.com/site/2011/02/08/greenpeaces-anti-forestry-campaign-re-emerges-at-james-cook-university/#more-1348.

ITS Global, 2011b. 'New Study by ITS Global Shows Palm Oil Can Spur Economic Growth in PNG.' Media Release, 22 March. Viewed 24 March 2011 at rhpng.com. pg/PDF_files/PALM%20OIL%20REPORT%20MEDIA%20RELEASE%20 FINAL%20V1.pdf.

IUCN (International Union for Conservation of Nature), 2008. *REDD Opportunities: Integrating Sustainable Forest Management Approaches*. Gland (Switzerland): IUCN.

John, M., 2011. 'An Open Letter to Rimbunan Hijau.' PNG Exposed Blog, 10 October. Viewed 10 October 2011 at pngexposed.wordpress. com/2011/10/10/an-open-letter-to-rimbunan-hijau/.

Kant, S. and S. Lee, 2004. 'A Social Choice Approach to Sustainable Forest Management: An Analysis of Multiple Forest Values in Northwestern Ontario.' *Forest Policy and Economics* 6(3/4): 215– 227.

Karousakis, K. and J. Corfee-Morlot, 2007. *Financing Mechanisms to Reduce Emissions from Deforestation: Issues in Design and Implementation*. Paris: OECD and International Energy Agency (COM/ENV/EPOC/IEA/SLT(2007)7). Viewed 12 January 2008 at oecd.org/env/cc/39725582.pdf.

Kellow, A., 2007. *Science and Public Policy*. Cheltenham (UK): Edward Elgar.

Lang, C., 2009. 'Vested Interests: Industrial Logging and Carbon in Tropical Forests.' REDD Monitor, 26 June. Viewed 28 June 2009 at redd-monitor. org/2009/06/26/vested-interests-industrial-logging-and-carbon-in-tropical-forests/.

Lattas, A., 2011. 'Logging, Violence and Pleasure: Neoliberalism, Civil Society and Corporate Governance in West New Britain.' *Oceania* 81(1): 88–107.

Laurance, W.F., T.E. Lovejoy, G. Prance, P.R. Ehrlich, G. Mace, P.H. Raven, S.M. Cheyne, C.J.A. Bradshaw, O.R. Masera, G. Fredriksson, B.W. Brook and L.P. Koh, 2010. 'An Open Letter about Scientific Credibility and the Conservation of Tropical Forests.' Scribd, 25 October. Viewed 26 October 2010 at scribd.com/doc/40046525/An-Open-Letter-about-Scientific-Credibility-and-the-Conservation-of-Tropical-Forests.

Lohmann, L., 2006. *Carbon Trading: A Critical Conversation on Climate Change, Privatization and Power*. Uppsala: Dag Hammarskjold Foundation.

Lutken, S.E. and A. Michaelowa, 2008. *Corporate Strategies and the Clean Development Mechanism: Developing Country Financing for Developed Country Commitments?* Cheltenham (UK): Edward Elgar.

Maguire, R., 2010. The International Regulation of Sustainable Forest Management: Doctrinal Concepts, Governing Institutions and Implementation. Brisbane: Queensland University of Technology (PhD thesis). Viewed 27 October 2011 at http://eprints.qut.edu.au/41688/1/Rowena_Maguire_Thesis.pdf

Miles, L. and V. Kapos, 2008. 'Reducing Greenhouse Gas Emissions from Deforestation and Forest Degradation: Global Land-Use Implications.' *Science* 320: 1454–5.

MNRE (Ministry of Natural Resources and Environment, Malaysia), 2007. 'Submission of Views by Malaysia.' 23 February. Viewed 11 June 2015 at unfccc.int/files/land_use_and_climate_change/lulucf/application/pdf/submission_by_malaysia_23feb07.pdf.

Mol, A.P.J., 1995. *The Refinement of Production*. Utrecht: Van Arkel.

Nabuurs, G., O. Masera, K. Andrasko, P. Benitez-Ponce, R. Boer, M. Dutschke, E. Elsiddig, J. Ford-Robertson, P. Frumhoff, T. Karjalainen, O. Krankina, W. Kurz, M. Matsumoto, W. Oyhantcabal, N. Ravindranath, M. Sanz Sanchex and X. Zhang, 2007. 'Forestry.' In B. Metz, O. Davidson, P. Bosch, R. Dave and L. Meyer (eds), *Climate Change 2007, Contribution of Working Group III to the Fourth Assessment Report of the Intergovernmental Panel on Climate Change*. Cambridge: Cambridge University Press.

Nelson P.N., J.A. Gabriel, C. Filer, M. Banabas, J.A. Sayer, G.N. Curry, G. Koczberski and O. Venter, 2014. 'Oil Palm and Deforestation in Papua New Guinea.' *Conservation Letters*. 7(3): 188–95.

Nelson P.N., M.J. Webb, I. Orrell, H. van Rees, M. Banabas, S. Berthelsen, M. Sheaves, F. Bakani, O. Pukam, M. Hoare, W. Griffiths, G. King, P. Carberry, R. Pipai, A. McNeill, P. Meekers, S. Lord, J. Butler, T. Pattison, J. Armour and C. Dewhurst, 2010. *Environmental Sustainability of Oil Palm Cultivation in Papua New Guinea*. Canberra: Australian Centre for International Agricultural Research (ACIAR Technical Report 75).

OCCD (Office of Climate Change and Development), 2013. 'Climate Compatible Development Policy 2013–2015.' OCCD, 12 February.

Oxley, A., 2009. 'The Paradoxical Risk of 'Ethical' Investment.' Forbes Opinions, 6 October. Viewed 11 June 2015 at www.forbes.com/2009/10/06/palm-oil-whole-foods-climate-change-opinions-contributors-alan-oxley.html.

Post-Courier, 2011. 'Armed Police Denied Entry to Esperanza.' *Post-Courier*, 22–23 October.

RH PNG (Rimbunan Hijau PNG), 2009. 'Will Papua New Guinea Benefit from Carbon Trading?' *RH Group Newsletter*, Issue 13, January–April. Viewed 2 May 2009 at rhpng.com.pg/PDF_files/RH%20NIUSLETTER_VOL13.pdf.

RH PNG (Rimbunan Hijau PNG), 2010. 'Climate Change: One-World Government Conspiracy.' *RH Group Newsletter*, Issue 16, October 2009 – January 2010. Viewed 10 January 2010 at rhpng.com.pg/PDF_files/RH%20 NEWSLETTER%20ISSUE%2016.pdf.

RH PNG (Rimbunan Hijau PNG), n.d. 'A Report on Climatic Similarities & Species Site Conditions: A Prerequisite to Introducing Plantation Species to Wawoi-Guavi TRP Area and Elsewhere in Papua New Guinea.' Viewed 3 March 2010 at fiapng.com/PDF_files/Report%20on%20Reforestation.pdf.

Rheeney, A. 2011. 'Rimbunan Hijau Welcomes PNG Police Investigations.' PNG Perspective blog, 10 October. Viewed 10 October 2011 at pngperspective. webnode.com/news/rimbunan-hijau-welcomes-png-police-investigations/.

RSB (Rimbunan Sawit Berhad), 2009. *Annual Report 2009*. Sarawak: RSB. Viewed 2 June 2015 at rsb.com.my/financial_information.html.

RSB (Rimbunan Sawit Berhad), 2013. *Annual Report 2013*. Sarawak: RSB. Viewed 2 June 2015 at rsb.com.my/financial_information.html.

RSB (Rimbunan Sawit Berhad), 2014. *Annual Report 2014*. Sarawak: RSB. Viewed 15 June 2015 at rsb.com.my/financial_information.html.

Stern, N., 2006. 'Stern Review: The Economics of Climate Change.' Viewed 15 June 2015 at http://mudancasclimaticas.cptec.inpe.br/~rmclima/pdfs/ destaques/sternreview_report_complete.pdf.

Stiglitz, J.E., 2008. 'Heroes of the Environment 2008: Leaders and Visionaries. Kevin Conrad.' *Time*. Viewed 1 January 2011 at rainforestcoalition.org/ documents/HeroesofEnvironment--TimeMagazine.pdf.

Tiden, G., 2011. 'Rimbunan Hijau Ripping Off Landowners in East New Britain.' PNGExposed Blog, 12 November. Viewed 1 January 2013 at pngexposed. wordpress.com/2011/11/12/rimbunan-hijau-ripping-off-landowners-in-east-new-britain/.

Tiong Hiew King, 2008. Speech by Datuk Tan Sri Tiong Hiew King. World Forestry Centre, 19 October. Viewed 17 March 2009 at rhg.com.my/media/ pdf/world-forest_ch-eng.pdf.

Tsing, A.L., 2005. *Friction: An Ethnography of Global Connection*. Princeton (NJ): Princeton University Press.

UNFCCC (United Nations Framework Convention on Climate Change), 2007. 'Report of the Conference of the Parties on Its Thirteenth Session, Held in Bali from 3 to 15 December 2007.' FCCC/CP/2007/6/. COP 13. Viewed 2 January 2008 at unfccc.int/resource/docs/2007/cop13/eng/06a01.pdf.

UNFCCC (United Nations Framework Convention on Climate Change), 2008. 'What is the Clean Development Mechanism?' Viewed 3 March 2009 at cdm. unfccc.int/about/index.html.

UNFCCC (United Nations Framework Convention on Climate Change), 2010. 'Annual Report of the Joint Implementation Supervisory Committee to the Conference of the Parties Serving as the Meeting of the Parties to the Kyoto Protocol.' FCCC/KP/CMP/2010/9. Viewed 15 January 2010 at unfccc.int/resource/docs/2010/cmp6/eng/09.pdf.

UNFCCC (United Nations Framework Convention on Climate Change), 2011. 'Clean Development Mechanism Registry.' Viewed 17 March 2011 at cdm. unfccc.int/Registry/index.html.

Verchot, L.V. and E. Petkova, 2009. *The State of REDD Negotiations: Consensus Points, Options for Moving Forward and Research Needs to Support the Process*. Bogor, Indonesia: Center for International Forestry Research. Viewed 11 November 2009 at cifor.cgiar.org/nc/online-library/browse/view-publication/publication/2870.html.

Wainwright, R., S. Ozinga, K. Dooley and I. Leal, 2008. 'From Green Ideals to REDD Money ... A Brief History of Schemes to Save Forests for Their Carbon.' Avoiding Deforestation and Degradation — Briefing Note 02. FERN. fern.org/publications/briefing-note/green-ideals-redd-money.

Wood, M., 2011. 'Royalties, Premiums, and Credit/Debt in a PNG Forestry Concession.' Paper presented at 'Melanesian Livelihoods Symposium', James Cook University, 20 April.

World Growth, 2009. 'Forestry and Climate Change: Cancún Changes the Game.' Press Release, 15 December. Viewed 3 January 2010 at worldgrowth. org/2010/12/forestry-and-climate-change-cancun-changes-the-game/.

World Growth, 2010. 'REDD and Conservation: Avoiding the New Road to Serfdom.' A World Growth Report, December 2010. Viewed 3 February 2010 at worldgrowth.org/site/wp-content/uploads/2012/06/WG_REDD_Conservation_Report_12_10.pdf.

World Growth, 2011a. 'How REDD Will Impoverish the Developing World and Reduce Biodiversity: An Indonesian Case Study.' A World Growth Report, April 2011. Viewed 3 February 2012 at worldgrowth.org/site/wp-content/uploads/2012/06/WG_REDD_Indonesian_Case_Study_Report_3_11.pdf.

World Growth, 2011b. 'Trees before Poverty: The World Bank's Approach to Forestry and Climate Change.' A World Growth Report, September 2011. Viewed 3 February 2012 at worldgrowth.org/site/wp-content/uploads/2012/06/WG_Trees_Before_Poverty_Report_11_11.pdf.